DISCARDED

Regional Integration in Central America

Studies in the Economic and Social Development
of Latin America
under the general editorship of
Otto Feinstein,
Monteith College
Wayne State University
and
Rodolfo Stavenhagen
International Institute for Labour Studies, Geneva

Regional Integration in Central America

Isaac Cohen Orantes

Lexington Books
D.C. Heath and Company
Lexington, Massachusetts
Toronto London

Copyright © 1972 by D.C. Heath and Company.

All rights reserved. No part of this publication may be reproduced or transmitted in any form or by any means, electronic or mechanical, including photocopy, recording, or any information storage or retrieval system, without permission in writing from the publisher.

Published simultaneously in Canada

Printed in the United States of America

International Standard Book Number: 0-669-81448-2

Library of Congress Catalog Card Number: 70-179093

Table of Contents

	Preface	vii
	Introduction	ix
Chapter 1	**Central America after the Second World War**	1
	The Units	1
	Postwar Politics	3
	Intraregional Relations	5
	Politics	5
	Trade	7
	Communications	7
	Migration	10
Chapter 2	**The Economic Commission for Latin America (ECLA)**	13
	ECLA's Theses	13
	ECLA's Strategy	15
	ECLA's Hegemony (1951-1959)	21
Chapter 3	**The United States Government**	27
	The United States Terms for Cooperation	28
	United States Participation	31
	United States Influence on Regional Economic Integration	35
Chapter 4	**The Program of Economic Integration by Sectors of Activity**	41
	Trade	41
	Free Trade	42
	The Common External Tariff	45
	Industrialization	47

	Regional "Infrastructure"	49
	Road Transportation	49
	Air Transport	51
	Port Development	53
	Electric Power	53
	Telecommunications	53
	Agriculture	55
	Monetary Cooperation	56
	Regional Financing	59
	Planning	62
	Technology	62
	Public Administration	63
	Other Sectors	63
Chapter 5	**The Institutional Setting**	67
	Institutional Evolution	67
	Institutional Autonomy	73
	Institutional Financing	73
	Participation	75
	The Pattern of Outcome of the Decision Making Process	77
Chapter 6	**Conclusions**	83
	Notes	87
	Bibliography	107
	Index	123
	About the Author	127

Preface

My main concern in writing this study was to analyze the origins of Central American economic integration and to critically evaluate its results from the perspective of its impact on the participants' development as well as its contribution to the reunification of these countries.

A large body of literature exists on Central American integration, particularly the numerous works produced by the institutions responsible for putting it into practice. There is also a considerable number of studies done mostly by foreign observers who were interested in the elaboration of a theoretical framework for the analysis of regional integration.

My interests were different from those of the studies mentioned. I was more interested in the Central American process itself and in accomplishing a critical evaluation of its results. I have relied on the existent studies, as it is acknowledged throughout the text, to achieve an understanding of a complex reality that would enable me to undertake its evaluation. Mainly for this reason, the study draws from several disciplines such as diplomatic history, economics, and political science, particularly from recent efforts within the latter to build a theory of international organizations. This has provoked some criticisms because the study, instead of being restricted to the boundaries of any discipline, intentionally avoids remaining within any one of them. I am still convinced that this approach is the only one that permits the understanding of the reality, which is always interdisciplinary.

My debts to those who have made possible this study are too great to be acknowledged sufficiently. Several persons and institutions made it possible. Among the latter I want to thank the Universidad de San Carlos de Guatemala, the Commission Fédérale de Bourses pour Étudiants Étrangers of the Swiss government, the Organization of American States (OAS), and the Ford Foundation.

The personnel of the following libraries gave me invaluable assistance: the United Nations and the Colegio de México libraries in México City; the library of the Permanent Secretariat of the General Treaty of Central American Economic Integration (SIECA) in Guatemala City; and the library of the Central American Bank of Economic Integration in Tegucigalpa.

Many persons made comments on earlier versions of the manuscript. Among them I want to thank Professors Jacques Freymond and Jean Siotis from the Graduate Institute of International Studies in Geneva; Mrs. Shirley Quinn for her editorial assistance; and Professor Rudolfo Stavenhagen whose interest made possible its publication. Finally I want to thank my wife to whom I dedicate the study.

Introduction

In this study I will describe the origins of the Central American process of regional integration and attempt an evaluation of the results of integration from a political perspective.

The facts and conditions that surround the launching of the integration program are rather obscure, in large part because when the persons or institutions that participated in these events have tried to describe them, they have produced laudatory narratives that tend, for obvious reasons, to disguise reality (a problem that affects "official" literature in any field of contemporary history).[1] In addition to the clarification of the process' origins, it is also necessary to appraise its results, so as to place the whole experience in historical perspective in accordance with the dictum that it is necessary to know the past to understand the present.

The study begins with an examination of the general conditions existing in the area when the program was launched. The study will show that these conditions were not favorable to the establishment of a program of regional integration. This leads to an analysis of the role and influence of two external factors: the Secretariat of the United Nations' Economic Commission for Latin America (ECLA) and the United States government. Their influence is studied separately since they participated in two distinctly different periods of the program's evolution.

To evaluate the results achieved by the process of regional economic integration from a political perspective (this constitutes the second part of the study), we need to define what is understood by the political consequences of regional economic integration. Given the breadth of the subject, it is necessary to delimit these results to avoid falling into a very general description that would make of the study merely a formal enumeration of those measures already adopted, without looking at their effect on the participating countries.

Theoretical work in the field of regional integration is a part of the study of international relations that has experienced considerable development in the last decade. Hence this will not be an inventory of all these theoretical efforts but only an attempt to point out those propositions that were useful in answering the question: what political results are theoretically expected from a process of regional integration?

It has been affirmed that, under certain conditions, regional economic agreements can become politicized and terminate in the erosion of the limitations of the nation-state by supplanting it with a larger entity. The attention of theorists has not been directed toward this terminal situation but to the process by which this is likely to occur, to an analysis of the conditions under which the process of successful politicization will take place. Thus, "neofunctionalism,"[2] as this theoretical position is known, defines integration as "the process of transferring exclusive expectations from the nation-state to some larger entity."[3]

The notion of politicization is basic to an understanding of the neofunctional paradigm of regional integration. It is "a process whereby the controversiality of joint decision making goes up."[4] This will occur by means of a "spill-over" of the original commitments into more controversial sectors or areas. "Spill-over" is the procedural engine of politicization, and it is understood as

the process whereby members of an integration scheme—agreed on some collective goals for a variety of motives but unequally satisfied with their attainment of these goals—attempt to resolve their dissatisfaction either by resorting to collaboration in another, related sector (expanding the *scope* of the mutual commitment) or by intensifying their commitment to the original sector (increasing the *level* of mutual commitment) or both.[5]

Before mentioning the conditions for this to happen it should be pointed out that this paradigm was originally enunciated with the Western European integrative experience in mind. For this reason, when its applicability to less developed areas was studied—by Haas and Schmitter in their well known article on the Latin American Free Trade Association (LAFTA)[6]—it was necessary to amend it to take into consideration the evident differences in the two settings. These amendments are important here, given the underdeveloped condition of the five Central American countries, Costa Rica, El Salvador, Guatemala, Honduras, and Nicaragua.

The adaptation of the neofunctional propositions was necessary, in the words of one of their authors, because "Latin America has not reached the 'end of ideology.' "[7] And it was done by looking for "functional equivalents" of the conditions that have made for the relative success of the Western European experience. These "functional equivalents" are: first, instead of Monnet's pragmatic-functional approach, ECLA's doctrine of industrialization by import substitution; second, instead of the European Economic Community's (EEC) bargaining process, ECLA's reciprocity formula; third, lacking the spontaneous links established among technocrats and private interest groups in Western Europe, the Latin American *técnicos* should try to establish these links; and fourth, the technocratic competence and anonymity of Western Europe should be replaced in Latin America by the building of alliances with political parties to provide support for the *técnicos*' activities.[8]

These "functional equivalents" of the Western European, or postindustrial setting were to allow in the Latin American, or less developed setting the emergence of a decision-making style favorable to spill-over and politicization of economic integration. But in explaining the operation of this decision-making style, neofunctionalism became prescriptive and made an important ideological option. It recommended the adoption of a strategy for the modernization of these countries, baptized by Hirschman as "reformmongering"[9] or reformism, by which the changes required by modernization were expected to occur without a direct confrontation with the existing socioeconomic structures.

This strategy assumed that the Latin American countries were in a "transitional stage," characterized by

a process in which the interplay between subjects and rulers is being enormously expanded, in which subjects demand more and more and rulers are challenged to change the established polity so as to accommodate the demands without destroying the system. Specifically, this process implies the rapid growth of an ever more complex system of politically articulate groups clamoring for attention, legitimacy and inclusion in the polity.[10]

This was ideally the fertile ground in which the neo-functional propositions were likely to prosper. The question that follows is: how was the strategy to be applied in a process of regional economic integration?

The *técnicos* were charged with the task of applying the strategy. Given their central role, the neofunctionalists assumed that during the "transitional stage" they acquired a "strategic position in the process of change."[11] From here on out they not only had to address themselves to those groups that were emerging and clamoring for attention, but, since this "transitional stage" was not yet over, also to the traditional interest groups, or to "key oligarchical interests" as Haas and Schmitter describe them.[12] The message that they had for these groups was the one contained in the ECLA doctrine, that is, import substitution and industrialization could be the solution to the backwardness of their countries. The *técnicos* had to try to convince these groups, and also certain political parties, of the advantages that they would derive from the application of ECLA's prescription.

The adoption of ECLA's doctrine also entailed the application of the reciprocity formula. This formula consisted of assuring all the participants that they would derive benefits from the process of economic integration, and of assuring the relatively less developed nations that they could achieve a higher rate of growth which would permit them to catch up with the relatively more developed. The enforcement of the formula would allow the *técnicos* to operate as agents for the common interest because these ambitious goals demanded total solutions. However, because it was impossible to solve all the problems simultaneously, the participants would become dissatisfied, particularly the less developed ones; the *técnicos* should then use this disenchantment, by playing the dissatisfied against the satisfied, to obtain the approval of other more universal solutions. Through this process the *técnicos*—this is the crucial point—would increase their autonomy because, as they were the ones that proposed and applied the solutions, the larger the solutions the greater their need for freedom of action. This decision-making style, it was affirmed by the neofunctionalists, would permit the spill-over of economic integration so that politicization could take place in the Latin American setting.

Finally, to the "functional equivalents" and the decision-making style was the added requirement that "the political leaders must find constant ways to feed the flame of integration sentiment, to make it attractive to the poor and induce the wealthy to make the major sacrifices.... This is a task," Haas concludes, "of statesmanship rather than of political science."[13] Political scientists prescribe, the politicians and *técnicos* act.

But spill-over and politicization are not the only outcome the neofunctionalists predicted as possible. They warned that the process of economic integration could also become "the prisoner and victim of the functional equivalents that permitted the establishment of the union."[14] But this was only mentioned as an alternative outcome that interested them less than the successful possibility. However, this study will consider both alternatives in its examination of the Central American integrative experience.

These were not the results expected by those responsible for putting the program into practice. The Central Americans were relatively more modest. They viewed the program of economic integration as an instrument for the modernization of the member countries, and even if they did not dismiss completely the possibility of simultaneously building a larger political entity among them, they felt that this was a very long-term projection and depended on the more immediate and urgent goals of development and modernization. It should be noted that these questions are not necessarily exclusive; there is a certain degree of complementarity which makes it possible to study them together. For this reason, the second objective of this study is double-edged because it tries to weigh the results expected by theoreticians and practitioners against the actual accomplishments during the period under examination.

In pursuing the second objective, I observe the sectors and institutions involved in the program. The sectorial description tries to determine the scope of the program by observing how many sectors and how much of each sector was affected by economic integration. This provides an assessment of the process of spill-over and politicization, as well as of the relative importance of integrative measures to the overall process of modernizing the Central American societies. I then proceed to study the role of the regional institutions within the integration process and try to establish the degree to which the larger entity emerged. This step includes the observation of the *técnicos*' role by describing the way the institutional setting evolved, the degree of financial autonomy it enjoyed, the degree of participation the process allowed, and the decision-making process.

The last chapter gives the general conclusions that can be drawn from the study by characterizing the program of economic integration and by trying to determine the way in which the expectations of theoreticians and practitioners were being realized.

This characterization of the process is done by means of the notion of "high" and "low" costs,[15] which permits an evaluation of the relative importance of integrative measures. It points out the fact that the process as a whole, in its origins as well as in its functioning, was characterized by the constant avoidance of "high" costs for the participants. It also tries to show the degree to which this constant characteristic hindered the achievement of the neofunctional expectation of the emergence, through indirect measures, of a larger political entity, as well as the painless development of the member countries, expected by those responsible for enforcing the program.

It is also necessary to emphasize that the main concern of the study is to examine from a critical perspective the results accomplished by the program of economic integration in the region during the period under analysis. Thus, rather than seeking the elaboration of a model or the polishing of an existent one, the study tries to point out the limitations of the reformist option that the process implies. In fact, the only theoretical ambition that could be attributed to the study consists in the criticism that it advances against the ideological expectations of the neofunctional theoreticians, as well as against those of the Central American practitioners. For this reason it can be stated that this study is more concerned with reality than with theory.

A final word is necessary on the period covered by the study—1950-1968. It does not include the war that started between two of the participant countries— El Salvador and Honduras in the middle of 1969—since, at the moment of writing, the conflict had arrived at a precarious cease fire which made it difficult to discern its effects on intraregional relations. The war—that occurred despite the existence of economic integration—marked the end of a stage in intraregional relations. This stage constitutes the main concern of this study.

1
Central America after the Second World War

This chapter will try to describe the conditions existing in Central America when the program for regional economic integration was initiated in the beginning of the 1950s. The choice of indicators to be studied has followed the theoretical contributions of other authors to the analysis of "background conditions" conducive to integration.[1] My purpose is to apply them in the Central American context, making the adaptations required by the region's environment.

Three sets of indicators have been chosen to describe the conditions in the region at the time. Some problems developed because of the low accuracy of Central American statistics; consequently, the data mentioned should not be taken as exact, but as indicating trends which may permit a certain degree of comparison.

The indicators to be analyzed are: the size and strength of the individual countries, or their "functional similarity or difference";[2] their political characteristics, bearing in mind that "the dialogue of political units is a function of the classes or the men in power";[3] and the rate and intensity of interaction among the countries.

The Units

It has been affirmed that similarity in size and power among the participating units of an integration process is, among other factors, "extremely favorable to the rapid politicization of economic relationships."[4] In other words, that it is easier for countries that are roughly similar to undertake economic integration and to transform their economic links into a more intense form of union.

In the case of Central America, the countries of the Isthmus share a number of characteristics, such as a common historical background, the same language, and roughly the same culture. But these common characteristics have not proved in themselves strong enough to support the repeated attempts to build a federation, thus confirming what other authors have observed in other contexts:[5] "we cannot use some previous historical experience which involved the notion of community as an argument for assuming the natural and inevitable reemergence of this happy state of affairs."[6] In spite of these shared factors which seem to favor integration in the area, the Central American countries have in fact conducted more than a century of independent life. They have existed as five separate countries, with differentiation and the development of nationalism as a consequence.

After the Second World War, the countries of Central America experienced an intense period of prosperity, due to an unprecedented improvement in the prices of their main exports in the world market. This period affected the countries differently. While Costa Rica, Nicaragua, and to a smaller degree El Salvador exhibited the highest rates of growth in the period, in Guatemala and Honduras the rate of growth of their economies hardly equaled the rate of population growth. The impact of prosperity on the economy of each country also varied, because of the different stages of development of the export sector in each country. In Costa Rica, El Salvador, and Guatemala the export sector was more developed, and consequently, they benefitted more from the increase in the prices of coffee, which represented the bulk of their exports. Honduras, in effect, had no national export sector. Bananas constituted the largest proportion of its exports and, although prices were relatively better at the time, the exploitation of this product was in the hands of the well-known banana companies. Nicaragua—the country that exhibited the highest rate of growth—was only beginning to develop its export sector.

In all the countries, the unprecedented period of prosperity was not due to an increase in production, but to an increase in the prices of the products. ECLA has characterized this period as one of "functional improvement in the economy, and not as a deep change in the structure of the system."[7] Because of the special conditions of the world market after the war, Central American imports and exports were concentrated in the United States, which between 1948 and 1953 bought around 75 percent of total Central American exports,[8] and between 1946 and 1951, supplied 73 percent of the five countries total imports.[9] This situation augmented the already considerable presence of the United States in the area.

Some of the countries began to industrialize, mainly because of the relative scarcity of imports and their war-caused isolation from external competition. Costa Rica, El Salvador, and Guatemala established textile and food processing industries. The expansion of the internal markets, caused by the prosperity of the export sector, contributed to these first steps toward industrialization. Nevertheless, in 1950 manufacturing was far from being an important sector of these countries' economies, only representing approximately 10 percent of their gross domestic product (GDP) and employing from 8 to 14 percent of all their economically active population. Agriculture, in contrast, contributed almost 50 percent to their GDP and employed about two-thirds of the economically active population. There were some differences among the countries in this respect, Nicaragua and Honduras appearing slightly above the last regional average.[10]

It can be said, then, that although these countries shared the common condition of primary product exporting countries, this generalization should not be carried too far, because (1) not all of them depended on the same products and (2) their different degrees of development influenced the impact that the conditions of the world market had on their economies individually. For the relatively more developed countries, prosperity meant the beginning of industrialization and social reforms; for the relatively less developed countries, the impact of prosperity was not as deep.

The public sector also expanded with the improved trade position, due to the importance of import and export taxes which represented more than 60 percent of governmental revenues for all the countries as a whole in 1950. Again, there were differences among the countries; for instance in Nicaragua this proportion was more than 80 percent and in El Salvador more than 70 percent.[11] Import taxes were the single most important source of governmental income, representing more than 40 percent in all the countries.

The fact that exports were the most dynamic sector in the Central American economies had influenced the content and intensity of their mutual relations. Transport and communications had been developed mainly to link the zones of export production to the world market. By 1950 all the countries were relatively well connected with the world market, but intra-regional and internal communications were poorly developed.

Postwar Politics

Postwar prosperity was accompanied by political changes in almost all the countries of the Isthmus, again with more intensity in some countries than in others. An analysis of the main values shared by the ruling classes or elites will help us to assess the homogeneity or heterogeneity existing in the region when the program of economic integration was initiated, having in mind that "the range of particular sectors of inter-state relations which are affected by the integrative processes occurring in the international system is proportionate to the degree of homogeneity."[12]

The countries of the region differed considerably in their social configurations. At one extreme appeared Costa Rica, with a fairly homogeneous social structure and a relatively reasonable distribution of wealth. In El Salvador the unequal distribution of wealth appeared as the main obstacle to political change. Guatemala had similar characteristics, aggravated by the existence of a large Indian population, whose living conditions have changed little since the days of the Spanish colonization. Honduras and Nicaragua had not yet developed a national export sector, consequently, subsistence agriculture was their main characteristic. The banana companies established on the northern coast of Honduras constituted the classical example of the modernized enclave surrounded by a backward economy. In Nicaragua, the family in power was simultaneously the most important entrepreneur.

In all countries—less in Costa Rica—income and wealth were concentrated in a very small percentage of the population, mainly due to the form of land tenure. Almost two-thirds of the population could be considered rural, and most of the rest was concentrated in the capital city of each country. The agricultural sector could be divided into a dynamic sector producing for export and a backward sector producing crops for domestic markets. In the latter case the land was:

largely in peasant hands in small holdings, mostly in the highland areas. Of about 800,000 farms which existed in Central America in the mid-fifties, about 80 percent could be classed as 'minifundia,' or very small farms. With an average size of 2.4 hectares, they accounted for 13 percent of the area in farms.[13]

This system of land tenure, whereby the land devoted to profitable, export production was in the hands of a few large landlords, was the main determinant of the social structure. Even as recently as 1967, "the participation of the peasant sector, which comprises half the population, in the money economy is small, probably accounting for no more than 10-15 percent of monetary circulation."[14]

This was the setting of the Central American political scene. The great depression of the 1930s brought to most of the countries of Central America the establishment of severe military dictatorships, supported by the landed oligarchy. Economic stagnation aggravated the already miserable conditions of the peasants, who participated in social uprisings in some countries that were violently repressed. Stability and maintenance of the status quo became the main concerns of the "depression dictatorships" of the 1930s. This period of stability and repression came to an end with a series of revolutions, more violent in some countries than in others, carried principally by the urbanized sectors of these societies.

This urbanized bourgeoisie, influenced by the anti-Fascist propaganda of the Allies, seized power in Guatemala in 1944. In El Salvador a successful *coup d'état*, also in 1944, was enough to oust the dictator; in Honduras the dictator was succeeded after elections by a member of the same party in 1948; and even in peaceful Costa Rica, an armed uprising was organized in response to an attempt to annul the elections of 1948 that the faction in power had lost. Nicaragua was the only country where the dictatorship was not challenged, and where it has maintained control of the country by a dynastic succession from father to son.

One of the main consequences of these political changes in all the countries was a broadening of the ruling elites, which in turn, produced a new conception of governmental functions. In contrast to the negative and repressive role of the "depression dictatorships," the new elites thought that it was the government's main responsibility to overcome the poverty of the masses within a framework of democratic freedoms. Consequently, the public sector was organized to this end: central banks were created, public works were increased, industrial and agricultural development was promoted, labor and social security legislation were enacted, illiteracy and health began to occupy the attention of the governments, and so forth.

These new governmental responsibilities required the participation of *"técnicos"*[15] who had the knowledge and the skills to draw up and operate the new programs. Consequently, relatively young professionals, mainly economists trained abroad, started occupying key positions in the governments of almost all the countries.

These changes took place with different intensities in the various countries. In Guatemala the revolutionary regime tried to go farther and started threatening the landowners and the banana companies by introducing a program of agrarian reform and transport development. The government was overthrown by a combination of the landed oligarchy and the army, supported by the United States Government. In El Salvador, after 1948, the new participants in government tried to enforce what they called "a controlled revolution." And in Costa Rica, the leaders of the movement of 1948 considered themselves "practical socialists" and believers in a "mixed economy."[16]

This is not the place to evaluate the contribution of these regimes to the modernization of their countries. Here we are interested in the influence these changes had on the more limited question of economic integration.

It has already been mentioned that one of its main consequences was the broadening of the ruling elites and that the values that inspired them were similar in almost all the countries. Nationalism and economic development within a democratic framework were the main values, but these shared aspirations did not in themselves necessarily lead to integration, because they were mainly conceived as remedies in the national context. ECLA provided the link to these national aspirations. But before studying ECLA's participation, it is important to examine the effect of these political changes on intraregional relations.

Intraregional Relations

Politics

During the time of the "depression dictators," the governments were oriented to the maintenance of order and stability, so their behavior could be mutually predicted, and the relations among them were stable and without major alterations. During World War II, these countries subordinated their activities to the strategic requirements of the Allies, to whose camp they all rallied, following the United States. But the revolutions transformed the homogeneity of this period. The concept of legitimacy of government became the main point of difference among them.

Among the principal values that the revolutionary governments were trying to enforce was government based on universal suffrage and democratic elections. Guatemala, where the revolution took place first, undertook an active foreign policy, the main purpose of which was to free the countries of the area from the evils of dictatorship: a Caribbean Legion was formed, mainly of political exiles, to invade these countries and depose the dictators. These activities were supported by other elected governments such as the regime of Prío Socarrás of Cuba. This "bonapartism" on the part of the revolutionaries was carried out in actions against the regimes in the Dominican Republic, Nicaragua, and in successful support of the Costa Rican revolution. This heterogeneity, brought about by the revolutionary principle of legitimacy, introduced political instability to the area mainly because of the low mutual predictability of behavior among the governments of these countries.

Consequently, the postwar period was one of political turmoil, and frequent clashes characterized relations among the countries. Since these conflicts were considerable in number and differed in intensity, this study will concern itself with only those conflicts requiring the participation of a regional or international mechanism, assuming that other differences were solved through channels of mutual accommodation. Between 1948 and 1959, the Central American countries asked for the intervention of the Security Council of the United Nations, the International Court of Justice, and the Organization of American States (OAS) in settling some of their differences. In each case, the OAS intervened first, and only later were some of the questions submitted to other institutions.

During the period under analysis (1948-1959), of the six times that the mechanism provided by the Inter-American Treaty of Reciprocal Assistance[17] was applied in the whole Hemisphere, five were to solve Central American conflicts. The government of Nicaragua was involved in all five of the Central American "cases": in three of them it supported political exiles—twice against the new regime in Costa Rica (1948 and 1955), and once with Honduras against the new regime in Guatemala (1954)—one concerned a territorial dispute with Honduras (1957-1961); and only one concerned an attack from Costa Rica and Honduras (1959).

The only question submitted to the Security Council of the United Nations was the complaint by the government of Guatemala in 1954 that it was being invaded from Honduras and Nicaragua by political exiles. The problem was never fully discussed because the invaders succeeded in deposing the Arbenz regime before a regional or international organization could study the matter.

The International Court of Justice dealt with one Central American conflict: the territorial dispute between the governments of Nicaragua and Honduras, caused by Nicaragua's refusal to accept the validity of an arbitrary award by the King of Spain in 1906. The Court, in November 1960, settled the matter definitively, declaring that the award of the King of Spain "was valid and that Nicaragua was obliged to execute it."[18]

These were the most important conflicts that affected the relations among the Central American countries in the decade during which the program of economic integration was initiated. As evidence that the main cause underlying almost all the conflicts of this period was the concept of legitimacy of government, it is interesting that Nicaragua, the only country where the dictatorship had not been ousted, was involved in them all.

Costa Rica was the main target of Nicaragua's interventions because the presence of democratic governments in Costa Rica and their demonstration effect had always been a source of concern for the Nicaraguan dynastic dictatorship. The Guatemalan crisis of 1954 was also a question of legitimacy, although the differences in this case were related to the ideological anxieties of the cold war. It involved the other four countries against Guatemalan regime, but only two—Honduras and Nicaragua—allowed their territories to be used as bases of departure for the invasion that brought back ideological homogeneity to the area.

El Salvador was the only country that was not involved directly during this period in a Central American conflict which had to be taken to a regional or international institution to be settled. Because it is overpopulated, it was frequently involved in migratory problems with its neighbors, but these were settled bilaterally.

The intensity of conflict in the region was not in direct proportion to the levels of other transactions among the units. Actually other types of interaction such as trade, communications, and migration were practically nonexistent, and no mechanisms existed for their development.

Trade

Trade among the countries was insignificant. In 1950 it amounted to $8.3 million or 3.6 percent of the five countries imports from third countries. Furthermore, more than 80 percent of this intraregional trade took place among Guatemala, El Salvador, and Honduras.[19] El Salvador appeared as the most important consumer of Central American goods, mainly agricultural surpluses from Honduras and some from Guatemala.

This relatively higher amount of trade among the northern countries can be explained by the existence of relatively better communications among them, and also because El Salvador and Honduras had had a free trade treaty for a limited list of products since 1916. Nevertheless, the amount of trade among the units was low by all standards, mostly because they all specialized in producing primary products for the world market.

Communications

The region's mountainous topography, added to the concentration of the population in the central highlands and the Pacific coastal strip, made any kind of transportation and communications extremely difficult and expensive. The countries had undergone the establishment of railroads and seaports at the beginning of the century to link the zones of production with the world market. Railroad was the basic means of communication among these countries and the world market. By the end of 1951, there existed 5,179 kilometers of railroads in the area, of which 2,899 kilometers, or 56 percent, were owned and used exclusively by the banana companies; of the 2,280 kilometers, or 44 percent, given over to public traffic, 1,464 were also owned by banana companies, and the last 816 kilometers belonged to private enterprises or to the governments.[20]

The railroads were linked exclusively to those ports which were served mainly by ships owned or controlled by the banana companies.[21] Of the five, Guatemala and El Salvador were the only Central American countries connected by a railroad; but its purpose was to connect El Salvador to a Guatemalan port on the Atlantic. Both the railroad and the port were owned and controlled by the United Fruit Company.

Road transport among the countries was almost nonexistent. In addition to the unfavorable topography, heavy rains made its development difficult and expensive. Few roads were all-weather, and paved roads represented a very small proportion of the existing total. The national road networks, with the exception of El Salvador, were inadequate and in all the countries served the most populated areas. Given these conditions, the development of regional road construction was given a low priority and attention was mainly focused on improving the situation in each country.

Nevertheless, an effort to connect these countries by road was begun in 1941, when the decision to build the Inter-American Highway was made. It is interesting to note that lack of access to the Panama Canal, by road, was one of the main reasons, and the United States agreed to pay for two-thirds of the total cost in each country. At the same time, the United States War Department started its own road project with the same purpose, but it was "badly conceived in its aims, since it did not help the war effort, and very badly executed, giving excess profits to the United States contractor," said an investigating committee from the United States Congress, adding that "only one third of the work done by the Army engineers is estimated to be useful to the future Inter-American Highway."[22]

By 1951 only 33 percent of the projected 2,018 kilometers of the Inter-American Highway was paved, only 44 percent of the roadbed was laid, and 23 percent was not built. Moreover, its completion was hampered by the small effort that the governments concerned were making. In the preceding five years, governmental expenditures budgeted for road construction went from 4 to 8 percent of their national budgets,[23] but with pressing internal needs, most of these funds were deflected.

Under these conditions, freight and passengers moved slowly and inefficiently from one country to another. In 1951, El Salvador—the largest consumer and supplier of Central American goods—imported ten thousand tons and exported six thousand tons by road, and of the twenty buses that could transport passengers from Guatemala to El Salvador, only one did it regularly.[24]

Transport by sea was not very important and its development was hampered by the fact that most of the population was concentrated in the Highlands and the Pacific Coast, where few natural ports existed. Even so, in 1951, 30 to 35 percent of the total regional trade was transported by sea—almost fifty thousand tons, including about twenty thousand tons of cement imported from Panama to El Salvador.[25] Intraregional sea transport was concentrated in the Gulf of Honduras, in the Atlantic, between Guatemala and Honduras, and in the Gulf of Fonseca in the Pacific, between El Salvador, Honduras and Nicaragua. Only two of these ports served the international trade of more than one country: Barrios in Guatemala's Atalantic Coast was linked to El Salvador by railroad, and Cutuco on El Salvador's Pacific Coast also served Honduras.

Air transport was introduced into the area in the 1920s. Because of the slow development of surface transportation and mountainous topography, it linked many isolated areas to the capitals—except in El Salvador—but it was a costly substitute for other internal means of transportation. Nevertheless, it was the only relatively efficient means of communication among the five countries, although its development had been disorderly and unsatisfactory.[26]

In 1951, the United Nations mission that studied the transportation situation in the Central American countries concluded that it "left much to be desired, its development has been unequal and unarticulated . . . , it is almost always inefficient, costly, and slow, and suffers from a remarkable absence of a national transport policy. . . ."[27] Under these conditions, the scarce attention that the governments were giving to transport in the area was hardly going to be oriented toward regional approaches, when the internal needs of each one of them were so pressing and so far from being satisfied.

Intraregional communications were not only hampered by the absence of the means to develop them, but also by the patterns of population settlement in each country.[28] These patterns, mainly influenced by climatic factors, show a clear differentiation between a northern and southern region for the area as a whole, with the southern showing a higher population density than the northern. Nevertheless, the most notorious differences in density and size of population appear in the countries individually. Almost two-thirds of the area's population was concentrated in Guatemala, El Salvador, and Honduras, and the other third was almost equally divided between the other two countries. But differences in the size of the countries are accompanied by differences in density. For instance, El Salvador, the country with the least territory, had the highest population density, and Nicaragua, the largest in size, was the country with the least population.

In each country two-thirds of the population was rural and the other third was concentrated in the capital cities. And because the urban centers are scattered throughout the area, communications among them were very difficult to develop. This fact is more easily appreciated if one considers that the Inter-American Highway—which connects the five capital cities—has a length of 2,003 kilometers from the Guatemalan border with Mexico to the Costa Rican border with Panama.

The main characteristic of the rural population of these countries was its dispersion. In 1950, 90 percent of the rural population of the region lived outside the limits of any administrative center, and less than 6 percent lived in settlements of one to two thousand inhabitants. If one studies their less populated zones, one can observe the differences in the pattern of population density from country to country. These zones—with densities of less than 15 inhabitants per square kilometer—occupied between 62 and 75 percent of the territories of Honduras, Nicaragua, and Costa Rica, and contained from 17 to 22 percent of their total population. In Guatemala, these zones occupied more than two-fifths of the total territory; but contained only 3 percent of its total population. El Salvador had no scarcely populated zones.

These patterns of population dispersion and concentration had resulted not only in a underutilization of land, but also in a low level of internal communications and, consequently, a lack of internal cohesion and integration. This lack was further intensified by the considerable differences in the distribtuion of income, especially as between the rural population and the urbanized sector. Finally, the dispersion of population affected the attitude of each country toward the possibility of regional integration. Thus, El Salvador—the only overpopulated country in the region—was the only country where integration was perceived by its national elites as a solution to a national problem and not as an idealized historical experience. El Salvador's favorable attitude toward integration translated itself into peaceful relations with its neighbors, bilateral free trade treaties, and a dynamic role in the development of the program of economic integration that was about to be initiated. None of the other countries exhibited such a real interest in integration and some of them, as in the case of Costa Rica, viewed it with less enthusiasm than the others, probably because its population was better integrated.

Migration

Migrations from one country to another within the region had been of minor significance. It has been established that immigrants proceeding from Central American countries represented, in the 1960s, less than 3 percent of the total regional population, with Costa Rica and Honduras being the largest recipient countries. This low index of mobility of persons has been characterized as follows:

Among countries that are not neighbors, migrations are of little importance and have been directed to the capitals;

Among neighboring countries, border migrations are generally the most important; and

In the countries where the most important migrations have occurred—from El Salvador to Honduras and from Nicaragua to Costa Rica—the most intense flows have been directed to the banana zones.[29]

Thus, the bulk of these movements consisted of seasonal agricultural workers and the rest were toward the capital cities. These last were important because in this way political exiles established contacts with the political elites of other countries in the region. Nevertheless, the movements of persons from one country to another were low and without much significance.

There were two fundamental obstacles to these movements. First, legislation which tended to discourage migration to control the political activities of exiles in the area, and second, the absence of considerable differences in salaries and labor conditions among countries, which could motivate movements of workers from one country to another. The few seasonal movements of workers that occurred support the second assertion because they were directed to the banana plantations where relatively better salaries were paid.

Finally, movements of tourists were also hampered by excessive formalities and controls.[30]

In 1950, only 40 percent of the 8 million inhabitants of the five countries knew how to read and write, ranging from 30 percent in Guatemala to 80 percent in Costa Rica.[31] In the same year, almost 300,000 newspaper copies circulated daily, there were 150,000 radio receivers and 180,000 cinema seats, and only 42,000 telephones existed in the whole area, most of these in the capital cities.[32]

Thus, intraregional communications were scarce, reflecting the low degree of development of internal communications. Some knowledge of each other existed among the urbanized sectors, but the bulk of the population—which was rural—had no contacts which could be considered conducive to regional integration.

This brief description of conditions in the region, indicates that, for the most part, the environment was not favorable to the success of integration. It is true that some favorable factors did exist in almost all the countries, such as the "developmentalism" that inspired the new ruling elites; but this condition was not complementary. On the contrary, it was mutually exclusive since it was perceived as a solution to the problems of each individual unit. Moreover, the heterogeneity caused by the political changes that brought the new elites to power was the most important source of conflict among the units, not a source of integration.

It becomes obvious then that economic integration in Central America did not stem exclusively from the internal conditions existing in the area. It had come through the influence of external factors.[33] Needless to say, the division among internal and external factors is made only for analytical purposes, since, as will be seen when discussing the influence of ECLA and the government of the United States, external factors cannot act by themselves—they depend on internal conditions to produce their effects.

2 The Economic Commission for Latin America (ECLA)

In February 1968 ECLA marked twenty years of activity with a resolution, approved by the member governments, recognizing, among other achievements, its "contribution to the scientific analysis of the economic reality"[1] of Latin America. Thus ECLA had achieved what it had originally seen as its first task in Latin America:

to analyze the social and economic reality so as to arrive at a diagnosis of the problems that faced its economic development and of the obstacles to the more persistent and accelerated growth of its economy. It was not an easy task to go against the tendency prevalent in the region, of trying to solve these problems with formulas that were applied in highly developed economies, where the social and economic reality was so different. Thus, ECLA initiated a current of thought adapted to an authentic Latin American reality.[2]

ECLA's first task consisted, then, of diffusing in the region a doctrine based on Latin America's reality. It attributed the problems of the area to external causes and the measures for their solution were prescribed accordingly.

ECLA's theses have been widely discussed in academic and international forums. Consequently, only a brief description of the ECLA diagnosis and prescriptions, and their impact on the Central American environment, with special emphasis on its integrative output, will be studied here.

ECLA's Theses

A division of the world into "industrialized centers" and "raw material producing periphery," places the Latin American countries in the latter. The main characteristic of the periphery is its dependence on the export of primary products, which is also its main problem: its persistent tendency toward external imbalance. The main reason for this trend is to be found in the deterioration of the terms of trade of Latin America's major exports due to deficiencies in the demand for raw materials and foodstuffs.

These are the problems diagnosed. The prescription consists of suggesting to the Latin American countries that they should free themselves from their dependence on the export sector by undertaking the production of manufactured goods. Or briefly, import substitution is the solution to the backwardness of Latin America. Or, to put it in Prebisch' words, "these countries no longer have an alternative between vigorous growth along those traditional lines and internal expansion through industrialization. Industrialization has become the most important means of expansion."[3]

Nevertheless, Latin America's heterogeneity has imposed on ECLA's analysis some differences in the way in which this prescription should be carried out. The industrial development of some Latin American countries is not only hampered by external conditions but also, among other internal conditions, by the smallness of their national markets. This was the case in the Central American countries. When ECLA started analyzing the possibilities of applying the prescription in this area, it found that the smallness of the national markets of the countries was the obstacle, even to the preliminary phase of easy import substitution, a phase which other Latin American countries had already passed through. Even so, Prebisch had already foreseen that this obstacle, "far from being insurmountable, ... is a factor which could be removed with mutual benefit by a wise policy of economic interdependence."[4] Therefore ECLA accompanied its proposals in Central America with a program of economic integration.

ECLA agreed that it was necessary to expand the national markets of the Central American countries by means of economic integration and that the possibilities of individual development of these countrries were unlikely. Economic integration was then a condition for the successful application of ECLA's prescription in Central America.

But these suggestions, even backed by the prestige that ECLA enjoyed in Latin America, were not enough in themselves. ECLA's proposal probably would have joined the list of unsuccessful attempts to reunite the area that characterize its political history, if it had not been for the political changes after the war which enlarged the political elites of most of the Central American countries. These political changes brought to key governmental positions what has been described as a "new generation, more concerned with the realities of economic growth than with the niceties of diplomacy or the rhetorics of political union."[5]

They were a generation of relatively young economists, most of them trained abroad, who thought that the main responsibility of government was to better the living conditions of the masses. But their thinking was limited by national boundaries. It was a search for individual solutions, without giving hardly any consideration to the possibility of a common one.

ECLA's message was addressed to these *técnicos*, but to gain their support it had to make clear that economic integration was not going to detract from their national efforts. On the contrary, ECLA showed them that the market expansion required by the industrialization of these countries would at the same time complement and ensure the success of their activities at the national level. This was the only approach available to ECLA to gain the support of this new generation, because if they had thought that regional economic integration would have detracted from their national goals they would have seen ECLA's plan as another empty reunification pledge among the many heard on the Isthmus.

ECLA's proposition was different from past attempts to reunite because its thrust was not to build a bigger nation but to overcome what the technocrats saw as the main problem of each country: underdevelopment. Taking the path toward reunification was justified only to the extent that it would contribute to an increase in the standard of living in each country.

It is almost certain that, without the external influence of ECLA, the possibility of transforming the region from a situation of mutual exclusiveness to one of integration and at the same time achieving each nation's goal of development could never had been considered. This statement can be corroborated only by examining the strategy with which ECLA accompanied its proposal for unification.

ECLA's Strategy

Three main features characterize ECLA's strategy in Central America: the separation of economics from politics; a gradual instead of total integration; and the carrying out of the program at a minimum cost to each country.

The first element is important because, by making integration an economic question, ECLA was freeing it from the political turmoil in which the countries found themselves after the war and also from any connection with the previous attempts at Central American unification. By presenting integration as a complement to the developmental aspirations of the technocrats, ECLA gave them a new justification for their presence in government. When their countries accepted ECLA's proposal, the offices created to implement the new programs—ministries of Economic Affairs, central banks, planning commissions, development agencies—were given new responsibilities which had the advantage of raising few controversies because this type of integration had general support.

The technocrats became the main channel of communication between the Commission's Secretariat and the governments because they were the members of the governmental delegations to the annual plenary sessions of the Commission. These delegations, mainly composed of the Ministers of Economic Affairs and Finance and their advisers, also had the unique opportunity of establishing contacts among themselves, with the members of the Secretariat, and with other international agencies. In this way the Central American "developmentalists" were introduced to postwar economic diplomacy.

During the third plenary session the Secretariat suggested to the member governments that, when formulating and adopting their programs for economic development, they should consider the possibilities of expanding demand by means of reciprocal exchanges, and thus achieving a better integration of their economies and an increase in productivity and real income.[6]

At the next plenary session, ECLA's Secretariat went on to propose the possibility of initiating a program of economic integration in Central America. The year between the two plenary meetings was one of preparation and contacts between the Secretariat and the governments concerned and led to the approval in México City in 1951 of resolution 9 (IV) initiating the program. This resolution invokes the "historical and geographical links that unite Central America," and states:

the interest of their governments in developing the agricultural and industrial production and the transportation systems of their respective countries, in a way that promotes the integration of their economies and the formation of wider markets, by means of the exchange of their products, the coordination of their development plans, and the creation of enterprises in which all or some of the countries have an interest.

It instructs the Secretariat "to study the measures or projects that permit the gradual realization of such purposes" and invites the governments "to proceed to the formation of a committee of economic cooperation, formed by the Ministers of Economic Affairs or by their delegates, that will act as coordinating body ... and consultative organ of the Executive Secretary of the Commission to orient the studies ... and consider their conclusions."[7]

Two important consequences derived from this resolution. First, it created the committee where the Ministers of Economic Affairs were to meet and second, it entrusted ECLA's Secretariat with the task of outlining the program to be followed. At the same meeting the creation by the Secretariat of a subregional office in México City for México, Central America, and some countries of the Caribbean was approved,[8] and four months later, Victor Urquidi was appointed head of this office.

Thus the guidelines of the program and its institutional setting were created. "It is to be noted," a prominent Central American official has said, "that the integration program was launched by means of a simple resolution, without fanfare and without the signature of any treaty."[9]

Simultaneously the Ministers of Foreign Affairs of El Salvador and Guatemala were trying to create a wider scheme of cooperation among the five countries to be known as the Organization of Central American States (ODECA). In October 1951, the Ministers of Foreign Affairs of the five countries signed the Charter of San Salvador by which a subregional organization within the framework of the United Nations and the Organization of American States (OAS) was created. It is important to examine the purposes of this new institution to note the differences between the approach toward unification of the economists from the development agencies and that of the lawyers of the Ministries of Foreign Affairs. ODECA's purposes were:

1. To strengthen the links that unite them
2. To devote themselves to maintaining the fraternal coexistence of this region of the Continent

3. To prevent and avoid all differences and assure the pacific solution of any conflict that might arise among them; to mutually assist each other
4. To search for joint solutions to their economic, social and cultural development by means of cooperative action.[10]

The texts speak for themselves; it is enough to add here that the generality that characterizes the latter has been one of the major causes for ODECA's failure, while the technocrats precision and more pragmatic goals have contributed to their relative success.

Among the institutions that the Charter of San Salvador established was an Economic Council under the supervision of the Ministers of Foreign Affairs, which submitted the activities of the economists to the scrutiny of the lawyers.[11] This situation was further aggravated by the creation of the Committee of Economic Cooperation where the Ministers of Economic Affairs would meet ex officio under ECLA.

But the *técnicos* foresaw the result of their submission to the *políticos*, and consequently, those who were part of their national delegations to the meeting who signed the Charter of San Salvador reacted immediately and obtained the approval of a resolution that granted them more freedom of action than was originally proposed in the Charter and reduced its submission to the Ministers of Foreign Affairs.[12]

Slowly, the Central American technocrats were freeing their activities from the politics and rhetoric of Central American reunification. This does not mean that they were not moving within the realm of politics, they were only doing it differently, because, as J.S. Nye has remarked, "a technocrat in Central America was less a pure technician than a new kind of politician whose style was that of the expert, and whose limited power arena was the bargaining room rather than the public square or military barracks."[13]

The separation among the activities of ODECA's Economic Council and ECLA's Committee of Economic Cooperation was emphasized when the first meeting of the Committee took place in Tegucigalpa in August 1952. Actually, two different meetings were held on that occasion: the Ministers of Economic Affairs, as members of the Committee of Economic Cooperation under ECLA, met for four days, and after this meeting was closed, they met for a day as members of the Economic Council under ODECA.[14] It was clear, then, that both institutions were to exist separately so that the Committee of Economic Cooperation could avoid involvement in the politics of the region.

Soon the fears of the technocrats were realized and justified. In April 1953, less than two years after the creation of ODECA, the government of El Salvador proposed the inclusion in the agenda of the first meeting of Ministers of Foreign Affairs of a point related to the infiltration of communism in the area. The Guatemalan government saw this as the formation of a political-military pact against it by the other four countries and withdrew from ODECA.[15]

In response, the Ministers of Foreign Affairs of the other four countries met in Managua to "reaffirm the democratic principles of Central America and condemn communism."[16] ODECA was resurrected only after ideological homogeneity was restored in the area, through an invasion of Guatemala from Honduras and Nicaragua, when the Ministers of Foreign Affairs of the five countries met in Antigua, Guatemala in 1955. They adopted, among other resolutions, one "enlarging the functions of ODECA's Economic Council," to permit it direct access to the governments to facilitate the accomplishment of its basic goals.[17]

Economic integration was now separated from political integration; the institution created under ECLA's auspices formed the nucleus for the program of economic integration. The technocrats were convinced of the advantages of such a setting because, in spite of withdrawing from ODECA, Guatemala had sent a delegation to the second meeting of the Committee of Economic Cooperation held in San José in October 1953. Thus, ECLA furnished a technical and noncontroversial setting for the program of economic integration.

While all these events were taking place, ECLA's México Office was studying the guiding principles and the concrete measures that should be adopted to launch the program. By the end of 1951, Prebisch, Urquidi and other ECLA officials had visited the five Central American countries and met with the Ministers of Economic Affairs, other governmental officials, and representatives from private enterprises and labor unions. The purpose of these visits was to discuss the proposals put forth by the Secretariat and to arrive at a preliminary consensus which would permit the launching of the program at the meeting of the Committee of Economic Cooperation in Tegucigalpa in August 1952.[18]

The proposals that ECLA's México Office advanced at that meeting outlined the basic principles of economic integration of the region and the fields in which immediate action was required to begin the program.[19] The proposals emphasized that integration was going to take place gradually and reciprocal industrialization was to be its foundation.

ECLA started describing the differences and similarities which existed among the countries, emphasizing those factors which were considered favorable to the development of integration, such as the interest of almost all the countries in economic development. Other factors discussed were: the demographic paradoxes of the region; the need for capital intensive investments to industrialize, and the weak rate of capitalization that the export sector permitted; the small size of the national markets as the main obstacle to industrialization; the actual improvement in the balance of payments situation of all the countries that had permitted the beginning of development planning; and the desirability of coordinating these plans to avoid investment duplication and to permit the specialization of industrial production. "These factors even without being decisive as arguments in support of an integration policy constitute elements that ... would now allow not only the paying of attention to the cyclical situation, but also to more long-term problems common to all the Central American countries."[20] The practical reasons that ECLA had in mind in dismissing the possibility of total immediate economic union were the effects of free trade on the inefficient industrial activities existing in the region, and the effects that a common external tariff would have on governmental revenues. For these reasons ECLA proposed to the governments "limited integration accompanied by a policy of commercial and industrial reciprocity."[21]

What did ECLA mean by limited integration with reciprocity? Gradual and limited integration meant a policy that "leads to the optimal location of some important economic activities ... with the purpose of establishing productive units of adequate size, that utilize the region's raw materials and are able to supply the Central American markets as a whole, at the lowest costs of production."[22] The advantage of gradual integration would be to establish new industries of adequate size, but instead of one for each country, there would only be one or two for the whole area. Gradual integration, at the same time, had to be reciprocally beneficial, because industrialization for ECLA was not "a process exclusive to a few only but, in a wide sense, it is a requisite of the development of all countries."[23] Thus, to avoid the repetition of the division of labor existing in the world—which ECLA severely criticized—in the more limited context of a regional scheme, it should be guaranteed for each participant that it would industrialize.[24]

Two major elements constitute the principle of reciprocity. First, industrial investments had to follow a general plan so as to avoid benefitting some countries more than others. Or as ECLA said to the Central American officials, "a general plan of establishing and enlarging industrial activities would have to be formulated for Central America, in such a way that some of the industries of optimal dimension included in the plan would be located in each country."[25] The second aspect of the principle of reciprocity was related to commercial and tariff policy and depended on the guarantee, given by all the countries, that only the products of the industries decided on in the general plan would enjoy the benefits of free trade within the area, "otherwise the maximum advantages of the optimal location of the activities would not be obtained."[26]

ECLA's proposal was also accompanied by a series of what it called requisites of the integration policy described above. The first was in the field of commercial policy and was related to the bilateral free trade treaties that El Salvador was entering into with the rest of the Central American countries[27] and to the practice of some of the countries of including a Central American clause of exception to the most favored nation provision in some treaties they had signed with third countries. ECLA considered the bilateral free trade treaties an intelligent approach, but believed that to be effective they should be linked to the policy of import substitution, that is, granting free trade to the products of the industries decided on in the general plan. A policy of free trade for the sake of increasing trade among the countries was not enough; what was needed was a commercial policy that encouraged the establishment of industries by granting them the exclusive privilege of supplying the expanded market for a limited period of time.

To put it in ECLA's own words, "the concept of commercial policy transcends its purely tariff aspects by requiring the adoption of formulas that guarantee access to the markets for which the industry has been planned."[28]

This first requisite confirms the assertion that ECLA saw economic integration as an instrument of reciprocal industrialization and not as an end in itself or as a device to obtain consensus.

The other requisite of the integration policy was related to questions such as the degree of protection that the industries would receive, the prevention of monopolies, and the development of regional services in fields such as technology, transportation, power, and financial cooperation.[29]

Finally, ECLA requested the governments to approve, within the general plan, a list of industries on which the Secretariat could start doing feasibility studies. This included cotton textiles, leather products, dairy products, forestries and pulp and paper, ceramics and glass, and so forth.[30]

The minutes of the meeting give the impression that the Secretariat's main function during it was to temper the overly ambitious proposals that the delegates advanced and to keep the program within manageable bounds. The outcome was a set of resolutions enabling the Secretariat to start on the first steps of its strategy.[31] The optimism of the delegates is best illustrated by part of the closing speech of the Minister of Finance of Honduras: "after a long century of tireless aspirations and efforts to perfect a plan that would again make us one, I think that now we have been able to discover the practical formula that permits to envisage such a goal."[32]

The last feature of ECLA's strategy relates to the costs of the program. By requiring that integration take place gradually, ECLA made clear to the governments that it was going to begin without hurting anybody. Gradual integration meant the adoption of a general plan to industrialize which had the virtue of assuring the participants that they would all get new industries. The only obligation that the adoption of this plan entailed was the reciprocal granting of free trade by all the countries to the industries that were to be located in each one of them. The integration program was also attractive to the governments, because by limiting it to industrialization they could be sure that they were adding something new to the modernization of their countries, instead of changing an existing situation as would have been the case if the program had started, for instance, with the agricultural sector. This would have meant attacking the traditional system of land tenure in each country and, consequently, a confrontation with the most traditional vested interests of their societies. For this reason the types of sacrifice demanded to the governments were in perfect accord with their attitudes toward social reform, that is, the transformation of their societies with a minimum of dislocation. This called for low costs and few changes. Industrialization was the sector that permitted such a policy, due mainly to its nonexistence in the area and the possibility that funds for its development could be obtained from outside the region. Higher levels of economic integration, such as a free trade area or a customs union, were dismissed not only because it was impossible to obtain the consensus for undertaking them, but also because these more advanced levels of integration could have negative effects on a few inefficient producers—a sufficient cause to consider them detrimental to the development process of the participant countries. The coincidence of a shared ideology among the local "developmentalists" and ECLA officials made the reciprocity formula not only a device for obtaining consensus, but because of their reformism, it was the essential means to foster the development of these countries through painless industrialization. A proposition that would have meant sacrifices for the national efforts or changes of a more qualitative nature in their societies would not have been adopted. Furthermore, the real costs of most of the studies and the experts to do them were going to be financed by the United Nations Technical Assistance Administration (UN-TAA). Thus, the low social, political, and financial costs of the program were among the most important characteristics of ECLA's proposal in Central America.

These three features of ECLA's strategy—separating economics from politics, gradual integration, and the low costs of the program—added to the way in which it would complement the "developmentalists' " national efforts, form the basis for the beginning of the program of economic integration in Central America. An analysis of the way in which ECLA went about applying the basic principles approved by the governments in the Tegucigalpa meeting will serve to illustrate that the first stage of Central American integration depended on the influence of an external factor on an internal situation.

ECLA's Hegemony (1951-1959)

The decisions adopted in the Tegucigalpa meeting authorized ECLA to undertake a series of studies. In accordance with the basic principle that Central American integration was not going to be a total process but an instrument for the planned industrialization of the participants, the first studies proposed were to be on: unification of international trade statistics and tariff nomenclatures; industrial technology and research; electric power; transportation; technical training in industry and administration; and financing of integration.

The next step was to decide on the level at which the discussions would take place. Since no immediate decisions were required because there were a series of technical details to be discussed first, it was decided to organize meetings at technical levels, where governmental experts could discuss the action required in related fields.[33] These technical meetings or subcommittee meetings, constituted the channels of communication between ECLA's México Office and the national administrations. Here the projects were prepared, which were afterwards submitted to the Committee of Economic Cooperation for final decision.

The first of these subcommittees was created to deal with questions related to the unification of tariff nomenclatures and foreign trade statistics. As time went on, whenever the Ministers decided on a new sector for the development of the integration program, a subcommittee was created to discuss and prepare the required project. In this way, during the period under analysis, subcommittees on trade, statistics, transport, electrification, and agriculture were established.[34] In addition, when a matter required technical analysis by the governments, *ad hoc* regional groups were organized, as in the case of industrial development and fiscal incentives.[35] These groups and subcommittees constituted a setting in which the program could be worked out in a noncontroversial way. Only once during the period did political events interrupt these activities. In 1954, with the crisis provoked by the ousting of the Arbenz regime in Guatemala, the annual meeting of the Ministers of Economic Affairs did not take place; only the trade subcommittee met in the second half of that year, when ideological homogeneity was restored in the Isthmus. Besides the regular meetings of the Ministers and subcommittees, two permanent institutions were set up: the Central American Institute of Industrial Technology and Research (ICAITI) and the Central American School of Public Administration (ESAPAC).

ECLA's México Office controlled these activities not only because it participated actively in the preparation of the projects but also because ultimately it held the purse strings. At the meeting of the United Nations Technical Assistance Board held in Geneva in September 1952, a group was created to handle the financial requirements of the regional program. This group was made up of representatives of the Technical Assistance Board, ECLA, FAO, the World Bank, and, when required, delegates from UNESCO and ILO. Also, a permanent regional representative of the Technical Assistance Board was appointed to supervise the requirements of the program. The requests for financial assistance were drawn up at the meetings of the Committee of Economic Cooperation, and after they were ratified by the governments, the chairman of the Committee sent them to the regional representative of the Technical Assistance Administration.[36]

The real costs of the program to the governments were low—in 1952-1953 governmental contributions amounted to $5,000 annually for each country, compared to $90,000 from the UN Technical Assistance Administration. The UN contribution increased annually and reached $300,000 in 1959, while those of the governments remained the same. In addition, UN grants represented almost the total budgets of ICAITI and ESAPAC.[37]

Thus, by providing the institutional setting, controlling the initiative, and securing the financial assistance required, ECLA's México Office fostered economic integration in Central America.

The achievements in this initial period are mainly in the field of commercial policy and road transport and are reflected in a series of treaties signed by the countries.[38]

When the report on the unification of tariff nomenclatures was submitted to the Ministers at the second meeting of the Committee held in San José in October 1953, the Nicaraguan delegation introduced the idea of free trade by proposing a multilateral free trade treaty. The proposal was dismissed by the other delegations as premature, but a resolution was approved authorizing the Secretariat to undertake studies in this field.[39] However, as free trade was not considered by ECLA an end in itself, it had to be linked to the establishment of the first industries in the region. For this reason, a series of industrial projects were approved, and an *ad hoc* group was formed to draw up the trade treaty.[40] These parallel activities produced in 1958 the Multilateral Free Trade Treaty and the Regime of Integration Industries.

The first followed the practice already common in the region of granting free trade to a limited list of products, with the commitment to expand the list by means of further "protocols" (treaties) so as to arrive at a free trade area in ten years, and pledged to form a customs union among the signatories "as soon as the conditions will be favorable."[41]

The other project put forward by the Secretariat was not approved as proposed. A comparison of ECLA's proposal with the text signed by the Ministers indicates that the governments were not ready to agree on a planned distribution of industrial activities. In other words, the role of the state in the economy was perceived differently by the governments and the Secretariat. The outcome—which supports this assertion—was a treaty outlining the principles on which the distribution of industries among the countries was going to take place, leaving to further "protocols" (treaties) such questions as the designation of the industrial plants to be distributed, the conditions for the establishment of similar plants, the measures for consumer protection, the participation of local and foreign capital, the protection to be granted to each industry, antitrust regulations, and so forth. The main argument used by the delegations for excluding these matters from the treaty was that "an excessive regulation could cause difficulties regarding the enforcement of its clauses, when applying the agreement to concrete cases."[42]

With the signing of these two instruments, the principle of planned distribution of industries—which linked free trade to import substitution and to the general norm of reciprocity that the Secretariat had sponsored from the beginning—was accepted. The next step consisted in establishing tariff equalization vis-à-vis third countries for the products included in the Multilateral Free Trade Treaty. These negotiations took place in the Trade Subcommittee, which submitted a project of treaty to the Ministers at the meeting of the Committee held in San José in September 1959. The treaty signed by the Ministers on that occasion established a common external tariff for a list of products, mainly those which already enjoyed free trade. Not all the tariffs were immediately equalized; in some cases, a period for adjustment was necessary. The mechanism to add new products to the list was similar to that used for free trade; that is, new products should be added by means of new protocols or treaties. In the same meeting another treaty was signed establishing a preference of 20 percent for the regional trade of those products not included in the free trade treaty. Finally, some products were completely excluded from these instruments for several reasons, such as the importance of their import duties to governmental revenues.[43]

The decisions arrived at in these two years consisted mainly of a treaty in which it was agreed to sign further treaties. That is, further steps on free trade, common external tariffs, and new industries were to be decided on a case by case basis. As in the launching of the Latin American Free Trade Association (LAFTA), this fact shows "there was no comprehensive consensus among the actors."[44]

Some local *técnicos* realized that the same painful process by which consensus had been obtained originally was going to be needed every time new measures were required. For this reason, disenchantment with the results thus far obtained started to be felt among the supporters of ECLA's activities in the region. The Multilateral Free Trade Treaty and the Regime of Integration Industries were considered "tiny, in the sense that, as instruments of integration, they fall short of the goal."[45] This malaise regarding the measures adopted was accompanied by a certain jealousy on the part of some technocrats toward the hegemony that ECLA's México Office had exercised for eight years. Furthermore, some governmental representatives questioned the lack of equilibrium between governmental projects and those of the private sector, a question related to the different conceptions that ECLA and the governments had of state intervention in the economy.[46] Other complaints were heard regarding the proposed distribution of the first industrial projects.[47]

Besides these complaints from those who supported the program, there was the fact that Honduras and Costa Rica had not ratified any of the treaties already signed. This meant that the program was developing at the pace imposed by the less enthusiastic governments.

Finally, the costs of the program were increasing and the United Nations Technical Assistance Board had to deny some requests because of lack of funds. For this reason, the Secretariat, at the meeting of the Committee held in San José in September 1959, suggested to the governments that they include some of the regional requirements in their national petitions for assistance, in other words, national projects would have to be reduced for the sake of the regional enterprise.[48] It was evident that under these conditions, the regional program would be sacrificed.

In addition to the reduction in financing, ECLA also planned to reduce its participation once the regional institutions created by the unratified treaties started functioning. At the same meeting, the Secretariat informed the governments that its participation would "tend to diminish at the product level, and it would instead concentrate only on those general studies requested by the governments."[49]

By 1959 the process was deadlocked, due mainly to the governments' lack of interest. For eight years these governments had been encouraged by ECLA to adopt and develop the regional program, but when ECLA started demanding higher levels of participation and sacrifices from them, such as increases in their financial contributions or the acceptance of free trade for a limited list of products, stagnation and disenchantment set in. However, at this very moment the process leaped into a more active stage, which can only be explained by analyzing the influence of another external factor: the government of the United States.

3 The United States Government

The main obstacle to integration in Central America was the unwillingness of the governments to place the regional program above their national preoccupations. But this attitude was not shared by those governments which perceived integration as a requisite to the solution of their national problems.

Probably the only country where integration was seen in this way by its national elites was El Salvador, where overpopulation made difficult to envisage exclusively internal solutions. But, at the same time, since the main cause of stagnation was the other participants' attitude, their cooperation would not be enough to overcome the deadlock. In this situation, it is not strange that El Salvador sought the assistance of the United States with this purpose.

Before describing the attitude of the United States government toward Central American integration, it is necessary to observe briefly the state of relations between the United States and Latin America at the time.

After 1945, United States economic policies were severely criticized by the Latin American governments. While the United States was seeking to enforce long-term, multilateral trade expansion within a framework of free enterprise, the Latin American governments were worried about the short-term problem of the effects that the removal of price controls in the United States would have on their main exports, on the dollar balances that they had accumulated during the war, and on their balances of payments. Furthermore, the Latin Americans felt neglected, because they thought that some of the attention that the United States was giving to the reconstruction of devastated areas should be given to Latin America too, since the immediate problems the region was going through were considered also a consequence of the war. This feeling led to a series of clashes over questions such as commodity agreements, regional development assistance, and even the more basic question of the creation of ECLA.

This period can be considered one of harassment and criticism of the United States in the field of economic policy. The main complaint of the Latin Americans was that the United States was indifferent to their efforts to protect their economies. Until the end of the 1950s they tried to get the United States to recognize, accept, and support their "developmental" efforts. This pressure was not a threat to the hegemony that the United States excercised in the Hemisphere —as it was understood in Washington—but rather its main purpose was to obtain material support for the development policies of the Latin American governments.

More than ten years passed before the United States saw these claims within the political context represented by the alternatives of reform or revolution, and not in the more theoretical or long-term context of free trade and free enterprise which characterized the beginning of the period. Only when these claims were presented as offering an alternative of controlled reform to violent and threatening revolution, did the United States start to support some of them. By the end of the Republican administration in the United States in 1958, clear evidences of a shift began to appear in its relations with Latin America. In that year, the United States supported the creation of a regional financial institution, an increase of quotas in the International Monetary Fund (IMF), and participated for the first time in the United Nations Committee on Commodity Trade. This shift in US policy reached its climax with the coming of the Democrats to power in 1961, when the reformist aspirations of most Latin American governments were shared by the United States government.[1]

Central America is no exception to the assertions made above; the strategic considerations which underlie the presence of the United States in the Isthmus—due in great measure to the Panama Canal in the South—have made it a traditional zone of North American interest. Central America did not start receiving development assistance from the United States in any considerable amounts, until the ousting of the Arbenz regime in Guatemala when the ideological anxieties of the cold war erupted in the area. Then, the United States started a bilateral assistance program with Guatemala to make of it "a showplace for democracy," and gave to the succeeding anticommunist government between 1955-1958 more than $80 million in grants and supported a loan of $18 million from the World Bank.[2] Never before had any country of the Isthmus received such large amounts of assistance. With the Cuban revolution and the establishment of the first socialist republic in Latin America in 1959, the United States interest in the Caribbean region became more intense.

All these events contributed to the shift in the attitude of the United States toward the development aspirations of the Latin American governments, and it began to support Central American integration.

**The United States Terms
for Cooperation**

The process through which the United States government came to support the Central American program of economic integration was a slow one. The fact that ECLA had sponsored it was a reason for early indifference. The United States had opposed the creation of ECLA in the United Nations Economic and Social Council (ECOSOC) meeting of 1947 arguing that it would merely duplicate the tasks of existing inter-American institutions.[3] And when at the meeting of ECLA held in México City in 1959—where the program of Central American integration was launched—it was decided unanimously to extend ECLA's functioning indefinitively, the US delegation reminded the participants that "its position respecting ECLA's continuation and terms of reference will be affected by overall rather than regional considerations."[4] Furthermore, ECLA's theses were considered heretic and in contradiction with the verbal commitment of the United States to free trade and free enterprise.

The United States regarded the Central American experiment with "cold indifference"[5] until ECLA advanced the proposal for the establishment of a Latin American common market and the question of economic integration appeared in the forefront of US-Latin American relations. Then the attitude of the United States changed from indifference to one of conditioned support.

At the meeting of the OAS Committee of Twenty One, held in Washington in February 1959, Undersecretary of State Douglas Dillon stated the conditions under which the United States government would support regional integration in Latin America. These six conditions are summarized here because they constitute the basis of United States participation in the Central American program:

Regional market arrangements should aim at trade creation and increased productivity through broadening opportunities for competitive trade and should not simply be trade diverting. This means that arrangements should provide for trade liberalization in all commodities—not just those in which members are competitive with non-members—and that duties and other restrictions applied by members of a regional market to non-members should not be higher or more restrictive after the formation of the market than before.[6]

It is evident that this first requisite contradicted ECLA's proposal to establish a preferential zone for the manufactured products of certain previously designated industries. Competitive trade was not ECLA's aim but protected free trade for certain industrial products—in effect a regional extension of the infant industry argument. This required higher degrees of protection to encourage the new industries, but the effects of this protection on nonmembers was not taken into consideration.

The arrangement should provide for a definitive schedule for the gradual elimination of virtually all barriers to intra-regional trade, and this process should be completed within a reasonable period of time. The United States does not favor an arrangement that provides simply for regional preferences with little more than a vague hope of eventually creating a free trade regime.[7]

This other condition contrasted with the Central American hope of arriving at a free trade area in ten years, by expanding a list of products established in one' treaty by means of further treaties whose ratification could not be guaranteed.

The arrangement should be in accordance with the principle of GATT (Article XXIV) for the creation of a free trade area or customs union and should be submitted to GATT approval. This is believed to be important not only because agreements of this kind must be reconciled with the General Agreement in order that the effectiveness of the GATT and the orderly system of world trade established under it may be preserved. If the Latin American countries were to set up their own arrangements without GATT's approval, it is feared that the whole GATT machinery might break down as a consequence of a proliferation of special regional preference systems all over the world. This is in accordance with the principle that regional arrangements should represent a step in the direction of world wide trade liberalization.[8]

This requirement was not new to the Central Americans; actually the Multilateral Free Trade Treaty had been submitted to the Contracting Parties of the General Agreement on Tariffs and Trade (GATT), due to the fact that Nicaragua was the only country of the five that was a member and had obtained waivers from the Contracting Parties for all the Central American Treaties which it had signed.[9]

Nevertheless, asking these countries to have in mind the consequences of their arrangement on world trade must have sounded somewhat bizarre to them. Because ECLA attributed the backwardness of the Isthmus to its dependence on the industrialized centers, it made the reduction of this dependence by means of inward-looking industrialization the main goal of the integration program. Worldwide considerations were at best secondary.

Regional trade arrangements should aim at increasing the degree of competition within the area. This means not only that virtually all commodities should be freed from restrictions on intra-regional trade, but that exclusive monopolistic privileges should not be given to particular industries or that there should be control agreements preventing competition. Not only is it believed that intra-regional competition will increase productivity and investment in the area, but that these conditions will also help to induce private investment.[10]

This requisite ran directly counter to the provisions of the Regime of Integration Industries which provided for the establishment of regional or "integration" industries that would enjoy exclusively the benefits of the expanded market for ten years. These were virtual monopolies, although they were accompanied by controls of quality and prices, their exclusive character was intended to operate not only as a protective device but also to attract new investment. The problem was that the United States did not mention how to avoid the possibility that investment would concentrate in the relatively more developed countries. Later this problem was acknowledged by the United States when they decided to create a regional financial institution to channel investment funds in the region.

Regional arrangements should provide not only for free trade in commodities but also for free flow of labor and capital in response to economic forces. Labor and capital should be free to move to places where they will be most productive. In this way it will be possible to achieve maximum benefits from economic integration.[11]

To the Central Americans this seemed a premature consideration. They had given little thought to labor or capital movements, mainly because the economic conditions of the countries were not very different. Also there was the danger that, like investment funds, these flows would benefit only the relatively more developed countries.

Any regional arrangement should provide for the financing of trade in convertible currencies. Neither bilateral payments nor a restrictive regional payments scheme which involves discrimination against non-members is justified.[12]

This last requisite was not much of an obstacle in Central America which had a long tradition of monetary stability; at the time, only Costa Rica and Nicaragua had limited exchange controls. However, if industrialization was going to be undertaken, these countries were going to have to raise protection and would start suffering the same balance of payments problems that were already endemic to the rest of Latin America.

From the above it is evident that having United States support meant a reordering of the program's main goals and the abandoning of some of the elements of the original scheme.

United States Participation

The turning point of the Central American program took place when President Lemus of El Salvador visited Washington in the spring of 1959, shortly after the clarification of the United States position on integration in Latin America.[13] Two questions related to Central American economic integration were discussed during the visit. The first was how to break the deadlock into which the program had fallen; this was mostly of interest to the government of El Salvador. And second, how to meet the requirements of the United States government.

The discussions of strictly economic matters were left to the Minister of Economic Affairs of El Salvador and the Assistant Secretary of State for Inter-American Affairs, at the time, Thomas Mann. The conversations were eased by the fact that Assistant Secretary Mann had until recently been United States Ambassador to El Salvador and therefore was personally acquainted with the members of the Salvadoran delegation and knew the current state of the integration program.

These officials, after acknowledging the "impasse" at which the program had arrived, concluded that a "real common market was needed."[14] The outcome was expressed in the joint communiqué, issued at the end of the visit, in which the Presidents said:

the establishment of an economically *sound system* for the integration of the economies of the Central American Republics and for a common market comprising those nations would be beneficial and would receive the support of El

Salvador and the United States... This subject will receive continued study by the two governments with a view to taking appropriate action to carry on those *sound plans* already contemplated (italics mine).[15]

Thus, first, a sound system was to be established. This term should be understood here as one that fulfilled the requirements of the United States. And second, the deadlock, due to the apathy of some of the members, would be surmounted by injections of foreign aid from the United States.

The next step was the visit to Central America of a mission of two experts from the State Department and the Vice-President of the Export-Import Bank "to consider prospects for helping the movement to advance."[16]

These activities led to the signature by El Salvador, Guatemala, and Honduras of what is known as the Tripartite Treaty, in Guatemala in February 1960. Under its terms these three countries accelerated the pace of the program.

This new treaty excluded the ECLA Secretariat and two of the original participants. It also differed in many respects from the other treaties signed by them all. For instance, free trade was accorded to all the products from the three countries, with a list of exceptions. Thus, the earlier decision to add products to a free trade list by means of further treaties was abandoned. The signatories agreed also to establish a common external tariff within a period of not more than five years. This was in accordance with the agreement for the equalization of import tariffs signed by the five countries in 1959, but contained the warning that if equalization did not take place in five years, the three northern countries would establish a common tariff among themselves. Finally, the three countries declared themselves in favor of the free mobility of persons and capital and of the maintenance of free convertibility of their currencies by stating that the persons and the investment coming from the other two countries would be treated as national in all of them. A fund of development and assistance to finance basic works and new industries, a directive council, an executive council, and a permanent secretariat were created to achieve these goals.[17]

One of the victims of the new scheme was the Regime of Integration Industries because, by granting free trade to all the products of the three countries, the exclusive right to free trade for ten years given to certain industries by the Regime could not continue. In this way the Tripartite Treaty not only accelerated the program but also changed the orientation of the program qualitatively, since rational industrialization—in ECLA's terms—ceased to be its main motivation. This question will be commented on later.

It is important first to summarize the main differences between the Tripartite Treaty and the original scheme: (1) the establishment of free trade was accelerated (it should be noted that the impatience of the local technocrats coincided with the US requirement of avoiding limited or very slow mechanisms of trade liberalization); (2) with the establishment of free trade for almost all the products, the program changed its main orientation from inward-looking industrialization to trade expansion; (3) the creation of a financial institution which has been discussed since the beginning of the program was assured with the participation of the United States; (4) monetary stability, which was hitherto included to prevent dislocations among the participants, was now included to meet the requirements of the United States; (5) free mobility of persons and capital was a completely new ingredient.

The step forward by the three countries could not have been taken without the support of the United States. Assistant Secretary Mann had said to the visitors from El Salvador: "we are willing to help in any way we can,"[18] which certainly meant that the Central American officials could count on complete support from the United States, as long as the program met the latter's requirements. Other observers of the Central American scene agree that "a strong case can be made that United States support was a vital condition for the important decisions that speeded up the integration program in 1960."[19] A prominent Central American official also supported this interpretation when he said that the measures were adopted "in the presence and not in the absence of an offer to meet their financial implications."[20]

This did not represent a change in the general attitudes of the participants. The importance of ECLA's assistance in the launching of the program also supports the assertion that economic integration in Central America was always induced from the outside or influenced decisively by external factors. By obtaining the support of the United States, the participants were only changing one source of assistance—the United Nations—for another—the United States—although this switch meant changing the general orientation of the program.

To explain this step forward by saying that it was induced by the most enthusiastic countries is not sufficient and moreover ignores the fact that Honduras—one of the three signers—had refused until then to ratify the treaties already signed, a fact that contrasts with the speed with which the Tripartite Treaty was ratified.

The signing of this treaty provoked strong reactions from the two countries that were excluded, Nicaragua and Costa Rica, and from ECLA's México Office. The most vociferous of the two countries was Nicaragua, whose exclusion was probably due to the fact that its territorial dispute with Honduras had not yet been resolved by the International Court of Justice, a problem which embittered its relations with its neighbor. Costa Rica, whose attitude toward the program had been isolationist and cautious (probably because it feared that some of its neighbors' internal political problems would be transmitted to it), saw in this situation an opportunity to divorce itself from the whole venture. The traditionally isolationist attitude of Costa Rica was reinforced at the time by the government of the "new conservative" Mario Echandi; with the accession of the opposition to power, it participated in the program.

ECLA's México Office reacted immediately, after it was informally notified of the changes that were going to take place by a member of the Salvadoran delegation to Washington. It tried to reconstruct the original five country program and to salvage some of its proposals which had been abandoned by the northern countries. It convoked an extraordinary meeting of the Committee of Economic Cooperation to discuss the situation in San José, April 26-29, 1960.[21] On April 27 the three northern countries deposited the ratification instruments of the Tripartite Treaty.

The flexibility with which the Secretariat handled the situation was remarkable. First of all, it did not consider that the acceleration in the formation of the free trade area decided by the three countries meant "a change in the nature of the objectives settled on previously by the five countries." On the contrary, ECLA thought that "it offered in a certain way a point of departure in the search for solutions to the problem,"[22] and advised the two outsiders to adjust to the pace imposed by the three insiders. The Secretariat concluded that the five governments should agree on establishing a common market among themselves in the next five years; thus the differences of time imposed by the Tripartite Treaty would only be transitory, since at the end of this period the five countries would arrive at the same goal. Furthermore, the adoption of this solution would permit the application of the Regime of Integration Industries among the five governments during the transition period, that is, before the establishment of full free trade in the area, since free trade "would make inoperative the main stimulus of the Regime unless it were possible to arrive at some type of agreement to preserve the effective guarantee of the Central American market to the integration industries, whose number, after all, would be small."[23]

Finally, with respect to the development fund created by the Tripartite Treaty, ECLA thought it advisable to organize a single development fund for two reasons: because the Secretariat had already been instructed by the Committee to study this possibility, and because the scarcity of financial resources among all the members demanded it.

ECLA's México Office did not see the crisis, then, as something insurmountable; on the contrary, it saw it as an opportunity to advance the original program. Nevertheless, the outcome was different from what the Secretariat suggested. Costa Rica was not convinced of the benefits of accelerating the program and, consequently, refused to participate in any attempt to do so. Nicaragua protested against its exclusion but decided to participate in the accelerated solution. With the withdrawal of Costa Rica,[24] the problem was practically solved, and the Secretariat obtained from the four countries the authorization to draw up a new treaty, instead of Nicaragua simply adhering to the Tripartite. The instructions given to ECLA's México Office were that the project should take into consideration the steps already taken by the three northern countries.[25] Negotiations for what is known today as the General Treaty of Central American Economic Integration started, and ECLA regained some of the ground it had lost with the signature of the Tripartite without its participation.

The terms of the General Treaty were negotiated in three meetings of the Trade Subcommittee held in the same year. A protocol to the agreement on tariff equalization which would add new products to the list was also discussed. Both treaties were finally signed by the four countries in the meeting of the Committee of Economic Cooperation held in Managua in December 1960.

At this same meeting the Ministers of Economic Affairs also signed a treaty creating the Central American Bank of Economic Integration (CABEI)[26] according to the guidelines developed in Washington. Shortly before the meeting in Managua, the State Department announced:

Discussions between the Ministers of Economic Affairs and other high officials of Guatemala, El Salvador, Honduras and Nicaragua and high officials of the United States Government, including the Assistant Secretary of State for Inter-American Affairs and the Assistant Secretary of the Treasury in charge of International Finance.[27]

The same communiqué added that the United States had decided to give "$10 million, $7 million upon the establishment of the Bank and $3 million in the next fiscal year."

For the member countries—excepting Costa Rica—this critical period resulted not only in an acceleration of economic integration, but also in the very important fact that the participants could count on the financial support of the United States to alleviate the sacrifices that this higher level of economic integration might demand from them.

United States Influence on Regional Economic Integration

With the arrival of the Democrats to power in Washington and the launching of the Alliance for Progress by President Kennedy in March 1961, United States policy toward Latin America changed. Its new orientation was to avoid revolutions in the area by emphasizing reform as a solution to the backwardness of the Latin American countries. Within this framework, economic integration was one of the most favored projects of the Alliance. This is not the place to undertake an evaluation of the Alliance, a question so widely discussed today; only its more precise contribution to the development of integration in Central America will be studied here, to illustrate the influence on its evolution of external factors.

Besides the favorable attitude of the United States toward reformism in Latin America generally, Central America acquired, due to the Cuban revolution, a significant importance in view of its strategic position in the Caribbean. The decision to depose the Cuban regime by supporting Cuban exiles was made by the government of the United States in March 1960. President Ydígoras authorized the CIA to establish training camps for the exiles in Guatemala, and President Somoza allowed the use of airstrips and Puerto Cabezas in Nicaragua as bases for the departure of the invasion that ended in the Bay of Pigs fiasco in April 1961.[28]

From then on the interest of the United States government in Central America increased, and most of the help given to the Central American integration program was inspired by this desire to avoid the spread of the Cuban precedent in the area.

The United States contribution to the integration of Central America should be studied in this context. Besides the opposition to ECLA and its theses, one of the first objections that the United States raised was to the Regime of Integration Industries. This instrument had three main objectives: (1) to encourage the establishment of industries of optimal size with exclusive access to the expanded market; (2) to avoid duplication of investment; and (3) to make industrialization reciprocally beneficial to all the participants. Compensating the relatively less developed countries to encourage balanced growth in the region.

In general the United States government opposed "integration industries" because "it considers that they will tend to limit competition and ultimately benefit neither the economy of the region nor the consumer. It is hoped that the regime of Central American integration industries is not a lasting feature of the regional economic integration program."[29]

In particular the objections were:

1. The scheme creates uncertainties which inhibit industrial investment and this, in turn, retards economic growth. . . . These problems are damaging Central American efforts to mobilize its own resources. The effect on potential investment from abroad is equally serious.

2. The process of designating integrated industries is arbitrary and involves the possibility of political favoritism.

3. The scheme would create monopolies. . . . The creation of monopolies by law or contract is undesirable.

4. The scheme is unnecessary. . . . The objectives sought through 'integrated industries' can be achieved through other means that already exist. One of these means, and one which the US Government regards as especially important is the 'decision of the market place'.[30]

This open opposition of the United States government to the Regime contributed to its failure—after ten years of being signed, only three industrial plants had been established under its regulations.[31] Nevertheless, although the Regime had become inoperative with the establishment of free trade for almost all products in the area, and although the objections of the United States meant that there would be no financing for industrial plants established under its auspices,[32] there still remained the question of balanced growth, theoretically one of the Regime's main objectives.

The problem of the equitable distribution of the benefits of the program was envisaged differently by the United States government. It was recognized that if the program was based exclusively on the forces of the market, the relatively more developed countries would obtain the largest benefits. Once the Regime of Integration Industries was doomed to failure, the United States manifested its predilection for financial means to solve the problems of balanced growth by proposing the organization of a regional bank.

In spite of numerous discussions among the Central American governments about the desirability of setting up a regional financial institution,[33] one must conclude that the step would not have been taken without the financial assistance of the United States—the first grant made by the United States to the regional program was to cover the expenses of establishing the bank.[34]

The preliminary discussions that led to the establishment of the bank were held in Washington, and from then on the United States maintained a close watch on the organization of the bank by appointing a "regional economic advisor for Central America"[35] who was simultaneously a "financial advisor" of the bank.[36] In early 1962 the regional economic advisor recommended in a confidential report that "his duties be amplified and taken over by a group of specialists."[37] In response to this suggestion a Regional Office for Central America and Panama Affairs (ROCAP) of the Agency for International Development (AID) was created in July 1962 thereby institutionalizing the influence and participation of the United States government. The position of ROCAP in the overall institutional chart of United States government agencies is, in a certain way, special, since it depends directly on the Director of AID. The fact that it does not have any institutional links with the United States' Embassies in any of the countries of the region, but only with the countries' AID missions, makes it appear a functional office occupied with the more precise question of economic integration in Central America.[38]

It can be assumed that in ROCAP the United States was trying to create an office similar to the one ECLA organized in México City at the beginning of the program, which allowed the program's evolution without controversy by sheltering the local technocrats behind ECLA's prestige. Needless to say, this "functional" aspect of ROCAP does not mean that it is divorced from the overall ideological objectives and security considerations that underlie the presence of the United States in the area.[39] A brief look at its objectives will help to understand the role ROCAP played in the development of Central American integration:

1. To strengthen regional integration institutions and their capabilities to formulate and execute regional development goals and plans

2. To promote mutual review and coordination (in Central America) of national development, fiscal, monetary, investment, and trade policies

3. To foster the further elimination of barriers of free movement within the region of goods, people, and capital

4. To improve private investment throughout the region

5. To reinforce bilateral US-AID missions efforts within each country to strengthen national economies and institutions as future components of a viable regional community[40]

With the institutionalization of US influence and participation through ROCAP, the program of Central American regional integration becomes one of the major achievements of the Alliance for Progress.

In the meantime Costa Rica was taking steps to join in the accelerated phase. Shortly after its government was changed by elections, Costa Rica, during the third extraordinary meeting of the Committee of Economic Cooperation held in San José in July 1962, adhered to the General Treaty, to the agreement creating the Central American Bank, and to the "Protocol of Managua" for the equalization of import tariffs.[41] The last step was taken when the bilateral negotiations relating to free trade and tariff equalization between Costa Rica and each one of the other four countries were concluded in November 1962.[42]

The original scheme was once again complete. In September 1962 the State Department announced President Kennedy's visit to Costa Rica to meet with the five Central American Presidents and the President of Panama.[43] The visit took place in March 1963 and constituted a public expression of United States approval and support of the integration program. The seven Presidents issued the "Declaration of Central America"[44] which indicated the orientation of the program for the immediate future.

The Central American Presidents expressed as their main goals: (1) the establishment of a customs union; (2) the coordination of their development plans; (3) the establishment of a monetary union and the unification of fiscal, economic, and social policies at a regional level; and (4) the enactment of the reforms required to modernize their countries, as stated in the Charter of Punta del Este, in such areas as agrarian, fiscal, educational, and public administration. Their ultimate goal was to achieve a Central American Economic Community.

The President of the United States offered to support the creation of a Central American fund for economic integration, as well as the development and financing of a regional plan and a regional housing program. All this financial assistance was to be channeled through the Central American Bank.

Finally, the Presidents declared that

in order to carry out their programs for social and economic betterment, it is essential to reinforce the measures to meet subversive agression originating in the focal points of Communist agitation which Soviet imperialism may maintain in Cuba or in any other place in America.[45]

Some of these measures were mentioned in the Declaration, such as supervision of travel to Cuba and a meeting among the Central American Ministers of the Interior to adopt common security measures to meet the subversive peril.

President Kennedy's visit had two clearly stated purposes. First, to support the integration program and second, to establish cooperative measures to arrest the influence of the Cuban Revolution in these countries. Shortly afterwards, Secretary of State Dean Rusk added another "point of importance" saying that the visit

had moved the Central American countries and Panama to the front of our attention as a nation and registered the fact that we look upon what is happening in Central America not as just part of the backyard or forgotten part of the Hemisphere.[46]

All the assistance offered by the United States was under the terms of the Alliance for Progress, where "self-help" was the requisite for its obtainment. For this reason, some time passed before the offers could be translated into actual loans and grants. Again the reluctance of the countries to put into practice further measures was the main source of delay. This reluctance was now aggravated by their unwillingness to fulfill some of the requirements of the United States, although this opposition should not be exaggerated. This was the case with the Regime of Integration Industries, which despite the evident objections of the US government, was incorporated—at the suggestion of ECLA—in the General Treaty of 1960.[47]

The same is true of the Central American fund proposed by President Kennedy in Costa Rica, which requires the adoption by the Central American countries of a permanent system of contributions to the fund. This situation was publicized in December 1964 when Thomas Mann, Assistant Secretary of State for Inter-American Affairs, said "much more could be achieved throughout the region as a whole if archaic doctrines about economic nationalism, protectionism and monopoly should be set aside in favor of true competition achieved on a gradual, step-by-step basis."[48] However, the countries did finally make their first contributions to the fund amounting to $10 million, and although no permanent system for contributing to the fund was forthcoming, the United States in July 1965 approved its first contribution of $35 million.[49]

Once more we see the importance of an external factor in the development of Central American integration. This assertion is particularly evident, when it is observed that further steps were taken by the participants only when they could obtain financial assistance to mitigate their sacrifices.

On the United States' side, its main interest in supporting the integration process was to preserve its traditional "sphere of influence"—even ECLA was considered "an uninvited meddler in Pan-American affairs."[50] But in rationalizing its hegemony, United States foreign policy—in the Isthmus, and throughout the Hemisphere—was characterized by a dualism that, on the one hand, invoked world responsibilities to justify its refusal to support the regional economic commission, and, on the other, did not hesitate to invoke the Monroe doctrine or the existence of the inter-American institutions to oppose ECLA's activities. This dualism was inspired by the will to isolate the Isthmus from extraneous influences, other than its own, and was characterized by the invocation of regional or universal responsibilities, as convenient.

The timing of United States participation in the Central American integration process shows that it took a threat such as the Cuban revolution to awaken its interest in the development and modernization of these countries. Containment and modernization were intimately related as the President of the United States declared in San José, in March 1963, "every nation present was determined that we should protect ourselves against immediate danger and go forward with the great work of constructing dynamic, progressive societies, immune to the false promises of communism."[51] But in face of the immediate danger, containment became more important than modernization, and the United States ended up supporting those forces that guaranteed the existence of things as they were and the fulfillment of its requirements, but that constitute the greatest obstacles to modernization. As Stanley Hoffman has said:

In Latin America the United States simultaneously makes efforts toward development and progress which cannot succeed unless they shake oligarchies and dislodge vested interests, and efforts to prevent subversion and insurgency which consist in rushing to the threatened gates and which therefore strengthen the *status quo*.[52]

The Central American integrationists were pragmatic and accepted the situation because it fitted in with their policy of avoiding costs and sacrifices by obtaining foreign assistance at any price. The question is how successful are these methods? The following two chapters try to analyze the process' results by economic sectors and institutions.

4 The Program of Economic Integration by Sectors of Activity

The program of economic integration will be analyzed in this chapter by studying the scope of integration—that is how many sectors of economic activity and how much of each sector is integrated. This analysis will lead to a study of the program's importance for the participants.

The sector approach is adopted to avoid the problems caused by other approaches, such as those concerned with "levels and thresholds."[1] This approach also provides the opportunity to study the dialectical evolution of the process—actions and reactions—in accordance with the opinion that "Integration and disintegration always occur simultaneously and that they are simply the two dialectical opposites of interactor relations in all international systems." Integration is here viewed as a process in which "consensus formation becomes the dominant characteristic of relations among actors in a system."[2]

This approach is particularly justified in the case of Central America since the process has taken place without a comprehensive instrument, in contrast to Western Europe. The Central American integration program was inspired by the general aim of promoting each country's own economic development, integrating gradually those sectors which, it was thought, would contribute to their painless industrialization. Consequently, we will deal with trade, the construction of a regional "infrastructure," industrialization, monetary and financial cooperation, technology, agriculture, and other sectors on a regional basis.

Trade

Initially, under the ECLA program, there was to be a preferential trade zone to encourage the creation of certain industries. With the participation of the United States and the signing of the General Treaty in 1960, the aims of the program were enlarged to include the establishment of a free trade area and a common external tariff vis-à-vis third countries for all products.

Before going further it should be noted that the Central American experience does not fit the traditional economic categories used to describe integrative efforts. The stages or levels of integration have been arranged here according to their increasing degree of controversy. Thus, in Central America, after the establishment of a free trade area—that is, granting free trade to all products originating in the member countries—comes the common market, which adds a common external tariff and the unification of commercial policy toward third countries.

The third stage is a customs union and adds to the previous two the free movement of all products within the area, a common customs administration, and the pooling of revenues derived from customs. The last stage is the economic union and entails, besides all the previous elements, the free movement of persons and capital and the establishment of common policies in economic and social fields. Following this schedule, the Central American program had reached the common market stage, and the members had decided to establish a customs union, which is considered a higher level of integration.[3]

Free Trade

By means of the across the board decision of 1960 which was incorporated in the General Treaty, the members granted free trade to all products originating in any one of them.[4] A list of exceptions contained two different types of products. These exceptions were agreed upon bilaterally to allow each member to exclude those products whose free trade would hurt existing producers, diminish fiscal revenues, or cause other major disturbance in their individual economies. The exceptions appear as Annex "A" of the General Treaty by pairs of countries.

For the first group of products, those allowed a five-year exception "to ease the adjustment of existing industries to new market conditions,"[5] the members settled on two requirements: first, the signature of specific protocols (treaties) regulating supply, as in the case of cereals; and second, the unification of import tariffs for assembled goods and their raw materials, as in the case of radios, automobiles, and wheat and wheat products. The importance of these products can be better appreciated when it is considered that in 1966 regional imports from outside the region of these goods, temporarily excepted from free trade, represented 16 percent of the five members' total imports, while intraregional trade in all liberalized products represented 17 percent of regional imports from abroad.

In 1966, 90 percent of nonliberalized imports from third countries fell into three categories of products: wheat and wheat products (11 percent), oil and oil derivatives (26.4 percent), and assembly goods (52.5 percent). Their exclusion from intraregional trade after the expiration of the five-year period reflects the unwillingness of the members to agree on their liberalization.

In the case of wheat, the main obstacle seems to have been the lack of agreement on a common tariff, due to Guatemala's fears of the consequences a relatively low common tariff would have on its producers of soft wheat. Furthermore, most of the countries had established plants for processing wheat flour from imported hard wheat.

The case of petroleum and its derivatives was in a way similar, because during the transitory period all the governments encouraged the establishment of oil refineries. The situation was further complicated by the importance that taxes on petroleum products had in the governmental revenues of all the members. Thus free trade in petroleum and its derivatives required agreement on tariffs and on producers of internal taxation. Finally, free trade in assembled goods awaited the signing of a specific treaty regulating the conditions under which plants could be established and coordinating procudures for internal taxation.

The distortions that this situation was causing in the region will be studied later on; here it is enough to point out that the slowness and difficulties that progress toward complete trade liberalization was facing were due to the perceived hardships that free trade could cause on the individual economies.

In the second group of exceptions—those definitively excluded from free trade—were found coffee, cheese, raw cotton, alcoholic beverages, matches, cigarettes. Also, in some cases that differ by pairs of countries, these products were quantitatively restricted. They were excluded from free trade for reasons similar to those that inspired the temporary restrictions, that is, because of their fiscal importance and to avoid regional competition.

Intraregional trade showed spectacular increases, from $21.7 million in 1960 to $260 million in 1968, representing 6.4 percent of the five countries' total imports in 1960 and 17.2 percent in 1966. This increase is frequently mentioned as an indicator of the program's success in this domain, and although it is certainly impressive, its importance in relation to total imports is minor.

Trade was originally concentrated in the northern countries. In 1960, Guatemala, El Salvador, and Honduras imported 80 percent of all regional exports; in 1965 this figure was down to 73 percent, and in 1968 down to 63 percent. This concentration of intraregional trade flows in the northern countries resulted from the fact that they represent 75 percent of the regional population and 70 percent of regional GDP, and also from the existence of more established trade flows among them and from the late participation of Costa Rica.

Not all the members shared equally in the expansion of intraregional trade. Guatemala and El Salvador persistently achieved surpluses in their regional trade balances, while Nicaragua had exhibited a deficit since 1961, Honduras since 1965, and Costa Rica since 1967. In 1968 these deficits amounted to $16.9 million for Honduras, $18.6 for Nicaragua, and $11.7 million for Costa Rica. Although it has been said that these deficits "are likely to reflect the earlier start of some countries toward industrial development rather than any fixed pattern resulting from substantial comparative advantage,"[6] they were the source of serious complaints.

The analysis of intraregional trade by commodities shows that manufactures predominated, representing 59 percent of all regional trade in 1965 compared to 40 percent in 1962. Also in 1965, intraregional trade in finished manufactures amounted to $79 million, while raw materials and foodstuffs increased in absolute terms but their share of the total trade dropped from 33 percent in 1962 to 17 percent in 1965. This can probably be explained by observing that the main thrust of the program was the industrialization of its members; agriculture was in a way neglected by trade liberalization, since agricultural products made up most of the exceptions. The program had still not overcome the previous geographical concentration of trade in agricultural products, and El Salvador continued appearing as the main importer and Honduras as the main exporter, followed by Guatemala.

There had also been intense trade among the countries of the same types of goods. This is seen by some as the beginning of product specialization by countries rather than interindustry specialization.

In spite of the increases in intraregional trade, not all the products originating in the member countries continuously enjoyed free trade for two main reasons: first, progress toward liberalization of the excepted items was slow and difficult; and second, the participants sometimes unilaterally hampered the free flow of goods already liberalized.

Finally, the decision to stop the products at the borders of the importing country is still in the hands of the national customs authorities, with the regional institutions intervening only after the merchandise has been stopped at the borders, on a case by case basis. This procedure caused considerable delays in deliveries.[7]

Free trade had been also hampered by the lack of attention given to the functioning of intraregional customs, probably because they are seen as one of the obstacles that the process will abolish. Meanwhile, customs authorities in all the countries exhibited a negative attitude toward free trade; "customs costs" represented more than one-third of transportation costs within the area; products in transit from one member country to another one traversed the territory of an intermediate country with a custodial agent, and this agent was paid by the exporter; sanitary controls—not unified among the members—were frequently used as excuses to hamper free trade. It has also been established that, since free trade meant, among other consequences, a reduction in fiscal revenues, some countries had imposed sales or consumption taxes instead, which were collected at the borders of entry of the importing country.[8]

The same reluctance toward free trade was exhibited by some Central American entrepreneurs, who had resorted to restrictive practices to mitigate the effects of regional competition by dividing the expanded market among themselves by products or lines of products. There had been no attempt by the governments or the regional authorities to enforce antitrust laws, which, in any case, did not even exist at a national level. To date, the only entrepreneurs in the region to have organized the marketing and distribution of products in accordance with the expanded market, "are subsidiary enterprises of foreign entities which count on a huge sales organization and a large financial capacity, or that, having traditionally exported to Central America, already have an established distribution system in each country."[9]

The questionable behavior of governments and local entrepreneurs indicated that free trade had not transcended the negotiation stage, and that its adoption had produced negative attitudes and reactions that neither the governments nor the regional authorities had been able to overcome.

The Common External Tariff

The decision to establish a common external tariff in Central America was adopted one year after free trade was granted to a limited list of products. In 1959, the member countries agreed (1) to equalize import duties for those products included in the free trade list, and (2) pledged to establish a Central American common tariff within five years. The first list adopted comprised the less controversial products, leaving the rest to be added by means of further treaties or protocols. The procedure has remained the same ever since, in contrast with the across the board decision of 1960 which granted free trade to almost all products originating in the member countries.

The period settled on in the 1959 agreement theoretically ended in September 1965 without the unification of the rest of the tariffs, because it was later decided that the five-year period should be counted from the date of entering into force of each subsequent protocol. Since 1959 six protocols had been signed which unified the tariffs for 97.8 percent of the unified customs nomenclatures. However, the 38 subitems still excluded represented 16 percent of the five countries imports from outside the area in 1966, with three products— wheat, petroleum, and assembly products—accounting for the bulk of these imports. The obstacles to the unification of their tariffs were related to their importance as fiscal revenues, and to the degree of protection that should be granted to existing as well as to new industries. ECLA saw these obstacles as: the problem of balanced growth, the survival potential of specific productive activities, and how regional sources would replace the goods traditionally imported.[10]

One of the consequences of the procedure of tariff unification was a rigidity of the tariffs already agreed upon, because their renegotiation required the same degree of consensus as the original decision. The negotiation procedure was characterized by its slowness. Each item was submitted to careful bargaining among governmental representatives, and once the tariff levels were agreed upon, a protocol was signed, leaving to each government the decision as to the right time to send it to the national legislatures for approval. This allowed the national administrations to delay the sending of the protocols to the legislatures and to postpone the deposit of the ratified ones in hopes of obtaining concessions from the other participants. Some regional authorities argued that the responsibilities for these postponements lay mainly in the legislatures, an explanation that obscures the fact that they occurred also in the negotiations, in sending them to the legislatures, and finally, in the deposit of the ratified treaties or protocols. The fact that—excepting Costa Rica—these countries were not the best examples of democratic tradition also contradicts this, since even during *de facto* regimes that legislate by decree, the process of national approval of the protocols was not accelerated. This question has been definitively clarified by ECLA:

the Central American experience indicates that the slowness in the renegotiation of custom duties does not lie in the congressional proceedings. Delays are also produced—and probably more prolonged—in other stages, and concretely: a) between the signature of the respective protocol and its presentation to the legislatures, and b) between the legislative ratification and the deposit of the corresponding instruments.[11]

The rigidity that characterized the Central American common tariff had great bearing on the more controversial question of industrial protection. The common tariff was looked upon in the region as an instrument of development, because it constituted the main device for fostering the industrialization of these countries. But, in view of its rigidity, individual producers sought other more flexible mechanisms to obtain protection from their respective governments. The governments, in turn, undertook a competition among themselves to attract the largest amounts of investment by means of national incentives or tax holidays. This competition, besides granting excessive protection to inefficient producers, hampered the establishment of free trade for certain products, did not permit the development of a common commercial policy vis-à-vis third countries, and finally, postponed the adoption of a common policy of industrial development based on sounder criterion than the competitive one.

This last question would seem to imply that the tariff levels agreed upon were considered insufficient to protect the industrialization of the members, and that for this reason, local producers sought other means to obtain higher levels of protection. The structure and height of the common tariff deny this implication, first, because it was inspired by a "selective policy aimed at modifying the composition of imports and to facilitate its substitution within a context of increasing foreign trade," and second, because under the common tariff the levels of protection in all categories were considerable—consumption goods, 85 percent, raw materials and intermediate goods, 34.4 percent, construction materials, 32.2 percent, and capital goods, 13 percent.[12]

Increases in the levels of protection are evident also when comparing the prior national tariff levels with those of the common tariff. Although no comprehensive study had been done on this question, the World Bank did a partial comparison and concluded that "the median of the list of ad valorem rates of duty of the previous tariffs was 44 percent but it is significantly higher, 64 percent, under the Common Tariff," adding that "the Common Tariff has resulted in widespread and often substantial increases in the rates of duties compared to both the simple or the weighted average of the previous national tariffs."[13]

Thus, it cannot be argued that the Common Tariff gave insufficient protection to local producers. On the contrary, it was the indiscriminate protection being granted to them, inspired by the competition among governments, which constituted the main obstacle to a tariff policy that would permit basing this sector on economic, rather than competitive critera. Because of the effect on inefficient producers and governmental revenues, the governments and regional institutions were unable to translate national aspirations into a common solution.

Industrialization

One of the main goals of the program of economic integration in Central America was the economic development of the member countries, which was understood as painless industrialization. As described earlier, ECLA's objective in this sector consisted of bringing the governments to agree on locating certain industries in each one of the member countries. For ten years the industries would have exclusive rights to the expanded market. To achieve these ends, the Central American Regime of Integration Industries was adopted in 1958. The treaty setting up the Regime contained the guidelines for the distribution of industrial activities within the area, but left the precise designation of each plant, its location, and the degree of protection and the controls it would require to the signing of further treaties or protocols on a case by case basis. That the Regime of Integration Industries was a failure is poignantly illustrated by the fact that, after almost ten years, only three plants were established under its auspices. This failure was due in part to the governments' attitude and, to a lesser degree, to the opposition of the United States government.[14]

It is interesting to observe that the main obstacle faced in the application of the Regime was the unwillingness of the participants to agree on the actual distribution of the industrial plants. While the decision was general, they could easily agree on its adoption, but when the moment to distribute the plants arrived, each country wanted at least one, otherwise it would not agree to the establishment of the others. In other words, each country saw the Regime as a means of obtaining a certain industry or industries on its territory and not as an instrument of a policy for regional industrialization. Thus, the creation of regional industries to satisfy an expanded market was hindered by the narrower question of the location of these industries.

This obstacle also relates to the equitable distribution of the benefits. The Regime's original objective was that each member would have at least one of the regional industries, and it assumed that in this way negotiations would be eased because each one would accept other plants in other territories in exchange for the location of one industry in its territory. Unfortunately the happy coincidence of five different projects that could be negotiated simultaneously to satisfy all the members failed to materialize; several members opted for the same type of industries. Even in the cases of those few industries already agreed upon, the establishment of a second plant to produce the same commodities showed that the Regime had failed as an instrument of coordination of regional industrial development.[15]

The Regime was also clearly opposed by the United States government which argued that the Regime encouraged the creation of monopolies in the area. This opposition meant, among other things, that no funds for "integration industries" were available from the main financial supporter of the program. Probably the Regime's faith would have been different if external sources of assistance would have existed.

The failure of the Regime prompted a search for other less inconvenient protectionist devices and for a method which would allow the national development of the manufacturing sector based on the regional market. One scheme was the Special System for the Promotion of Productive Activities, whose main purpose was to grant to selected activities a higher degree of protection than the one already allowed by the common external tariff, without any time limitation for the protection granted and without the controls imposed by the Regime of Integration Industries on prices and quality. In short, this device constituted a new procedure to obtain excessive degrees of protection. A further aggravation was the fact that individual producers made their petitions for tariff increases to their governments, who presented them to the regional institutions as national, not private, issues.[16]

Also related to industrialization was the persistence of the use of fiscal incentives to encourage industrial development at a national level. Competition among the members to attract the largest amount of investment by concessions of fiscal exemptions or tax holidays distorted investment decisions in the area because industries located where they could obtain the best deal. The agreement on the unification of fiscal incentives signed by all the members is illustrative. This agreement's entering into force was delayed because Honduras feared that its situation of relatively less developed country would be aggravated once fiscal incentives to industrialization were unified throughout the area.

Finally, the ratification of the agreement for the unification of fiscal incentives will not bring an end to this situation of national competition to attract investment. One article of the agreement leaves the concession of the unified incentives to the national administrations for a period of seven years, and only until then a regional institution will grant them in the area.[17]

The observation of the instruments that the members adopted for the industrialization of their economies within the regional program shows that—due to the participants' unwillingness to give up the incentives with which they used to attract the investment that the expanded market generated—a regional policy in this sector was not possible.

Despite the fact that manufacturing constituted the most privileged sector of the integration program, less than 20 percent of the five countries' industrial output was traded regionally between 1962 and 1967. Thus, the growth of 11 percent per annum experienced by the manufacturing sector in this period cannot be credited to the program of economic integration but rather to the increases in local demand due to the improvement in the export sector experienced by the five countries during the same period. It has been estimated that some "three quarters of the increment in manufacturing activity since 1962 was due to an increase in local or national consumption."[18]

The structure of the manufacturing sector was not substantially affected by the regional program, since, in the period under analysis, the traditional consumer goods industries—processed foods, beverages and tobacco, textiles, shoes and clothing—still accounted for 75 percent of regional industrial output and represented about one-half of the increase in regional exports of manufactured products, with recent increases in trade in nontraditional manufactures, such as fertilizers and petroleum products, being due "to certain factors which may not be of a permanent nature."[19]

The rest of the manufacturing sector was made up of industries that performed a very low proportion of value added within the area and that required low capital investment, such as pharmaceuticals and cosmetics. The appearance of these industries was largely attributable to competitive national fiscal incentives, rather than economic considerations.

The intention frequently expressed by the regional authorities of working out a coordinated policy for industrial development in the area, remained only an intention as long as the participant governments competed among themselves in attracting investment by means of national policies of exaggerated protection, which in turn encouraged inefficient industries and investment duplication. Given the scarce levels of investment in the region and the small size of the expanded market, these self-seeking practices were pernicious and, in the long-run, unwise.

Regional "Infrastructure"

When the program of economic integration was launched, ECLA and the Central American governments realized that it demanded the development of certain sectors which would increase the contacts among the members, and whose joint development would represent considerable reductions in costs. Consequently, ECLA suggested the adoption of regional measures in fields such as transportation, telecommunications, and electric power, to construct, what it called, a regional "infrastructure." This demanded financial resources, which the Central American countries, evidently, did not possess. Again they had to depend on external means to achieve their goals.

Road Transportation

One of the first studies done by UN experts in Central America was on the transportation needs of these countries. It was the first study done in the area on the individual transportation systems which considered the possibility of economic integration in drawing up its guidelines for future action. It emphasized the need to develop and link the national transportation systems and suggested that priority be given to road transport, which they considered the most efficient way to satisfy the internal needs of the region. Railway transportation devoted to the export sector of the national economies was relatively adequate for the time being. Emphasis was therefore placed on road transport and the internal needs of the Central American countries.[20]

Discussions were held among the national transportation authorities on how to implement these suggestions. These meetings led to the signing of two treaties unifying road signals and regulating road traffic throughout the region by the Ministers of Economic Affairs in the Committee of Economic Cooperation in 1958, and also to the creation of a transportation subcommittee under ECLA to coordinate national policies in this sector. One of the first tasks of the subcommittee was the standardization of road construction and design norms, but concrete regional actions could not be undertaken until foreign financing could be obtained. During President Kennedy's visit to San José, in March 1963, the United States government suggested the creation of a fund administered by the Central American Bank "to finance the improvement and construction of the Central American regional road network."[21]

One month earlier officials from SIECA, the Bank, and ECLA, actualized the study done by UN experts in 1953 and proposed the consideration of thirteen roads as of utmost regional importance. At a joint meeting of the Ministers of Economic Affairs and Public Works held in September 1963, it was agreed to construct the thirteen roads suggested previously, with a length of 1500 kilometers and at a cost of almost $75 million.[22]

Because the Inter-American Highway was almost built at the time, the regional network was to be based on this trunkroad with the following objectives: to increase the flow of traffic and improve the conditions of road transport in the region; to link the northern part of the Isthmus by means of interoceanic roads; to link zones of agricultural production to consumption centers; and to build a basic road network in Honduras.[23]

Construction of the regional network was projected in two stages, 1963-64 and 1965-69, but this was considered optimistic by some; "it is unlikely that more than 80 percent of what had been projected will be achieved by that time."[24]

The main source of financing for the regional road network was the Integration Fund of the Central American Bank whose resources amounted to $120 million in April 1969, of which slightly more than 80 percent consisted of loans from the US-Agency for International Development (AID) and the Inter-American Development Bank (IDB), and the rest were the participant's contributions. As of June 1968, the bank had committed loans for road construction that amounted to $58 million from the fund, these loans covered the total of the estimated project cost at 3.5 percent annual interest over a twenty-five year period (including a grace period of seven years). These terms explain the predilection exhibited by the governments for this source of financing.[25]

Despite the availability of funds and the agreement on the selection of the regional roads, there did not yet exist a regional criteria for determining the order in which the roads would be built; this decision was still in the hands of the governments. This has led regional observers to conclude that, from a regional point of view, the progress achieved in road building is not very important when compared to governmental activities to solve national problems in this field.[26]

At the same time two of the roads included in the plan were only important because they traversed national boundaries: "their benefits are doubtful, and their construction and betterment should be postponed until their economic justification clearly supports the expenditures involved."[27] Other regionally important roads have not been included in the plan.

The absence of regional priorities is probably more evident when observing the projections done by regional institutions and governments. The Central American Bank financed a study in 1964-1965 that identified a list of sixty-nine projects to be built in the next ten years at an estimated cost of $340 million. The Permanent Secretariat of the General Treaty (SIECA) had a plan that envisaged the construction of 9,648 kilometers of roads over a twenty year period at an estimated cost of $768 million. Finally, the five governments national plans revealed that they projected the construction of 5,120 kilometers of roads, within the regional program, at a cost of $120 million during 1965-1969.[28]

Road construction in the area had experienced spectacular increases, going from 10,230 kilometers of paved and all weather roads in 1953 to 24,816 kilometers of the same roads in 1963. Foreign assistance financed 52 percent of all road construction in the area during 1955-1963. For the following period, 1965-1969, projected investment in road construction by the national plans represented 30 percent of overall public investment, with foreign assistance financing about 75 percent of this effort.[29] Thus, it appeared that the members would continue to develop their national transportation systems within a framework of weak regional coordination and considerable external support. Meanwhile, regional transportation costs were higher than national ones for the same distances and regional tariffs and services were not organized to serve the expanded market.

Air Transport

The development of air transport in the region constituted a costly substitute for other means of transportation. Even so, there were parts of almost all the countries—except El Salvador—which could only be reached by air. Air transport was still the most efficient means of transportation among the five Central American capitals, and in these two senses has an important function in the economies of all members.

In this field the program exhibited one of its most interesting examples of regional cooperation in the form of a corporation composed of the Directors of Civil Aviation from the five countries and created by a treaty granting it "exclusive rights to provide air traffic services and radio aids to air navigation in the territories of the contracting parties."[30]

The preliminary discussions on the creation of this Central American Corporation of Air Navigation Services (COCESNA) took place among the Directors of Civil Aviation of the five countries in a meeting held under the initiative of the International Civil Aviation Organization (ICAO) in Tegucigalpa in 1957. Initially a regional flight information center was organized in Tegucigalpa. The fact that this center left the organization of services to each country, and in some cases to private airlines, led to excess capacity and waste. The Honduran delegation to the meeting of Directors of Civil Aviation held in Tegucigalpa in 1959 proposed the creation of a regional corporation, with ICAO's technical assistance, to avoid these faults. The COCESNA agreement was ratified by four countries in 1961—Costa Rica joined later.

COCESNA was financed by an initial contribution from each government of $20,000 and was furthered by users charges from the airlines, foreign assistance, and the nationally owned airport equipment and buildings that were transferred to COCESNA. The creation of COCESNA aroused some opposition from the airlines, especially from the only foreign one operating in the area at the time, which demanded, without success, that the Supreme Court of Honduras—where the Corporation had its headquarters—declare its existence unconstitutional. (The airline owned and operated its own equipment.)

After overcoming this initial obstacle, COCESNA went on to fulfill its main goal of improving and coordinating air navigation services, with considerable support from ICAO, US-AID, and the Federal Aviation Agency of the United States government. By leasing to the governments its idle capacity in radio channels and thereby making possible telephone communication among the five Central American capitals, it relieved a pressing need.[31]

It can be concluded that there were few hindrances to the development of COCESNA. This was probably due to the relatively small sacrifices that it demanded from the individual governments, given the financing, technical assistance, and equipment that COCESNA received from foreign sources.[32]

Other activities related to air transport did not progress as well. Airport development was undertaken to satisfy national needs without considering possible joint ventures. Another case of uncoordinated national activities, leading to waste and inefficency, was the existence of six national airlines in the area. This proliferation of enterprises was caused, among other reasons, by considerations of national prestige. But despite the evident advantages, as of 1968 a regional consortium had not gone beyond the stage of preliminary discussions, out of which came the conclusion: "to organize fewer but stronger airlines will require governmental initiative and an act of a high degree of international cooperation."[33] Both of these requirements were conspicuously absent in the area.

Port Development

The existing sea ports in the region served the participants' foreign trade needs and were not affected substantially by the program of economic integration. Nevertheless, a working group within ECLA's transportation subcommittee was organized and met twice, in November 1967 and September 1968, to discuss the unification of tariffs, statistics, and practices. Apart from these preliminary contacts, port development was based on national considerations without attention to intraregional trade or any other regional perspective. One expects little change in the near future since the five countries individually projected investments in this sector for the period 1966-1970 of approximately $50 million; these projections included, in some instances, the duplication of existing efficient ports located nearby in neighboring countries.[34]

Electric Power

In the period under study, national activities concentrated on hydroelectric development, given the absence of mineral sources of energy in the area, and joint efforts were limited to examine the possibilities of interconnecting pairs of countries. The obstacles that these projects of interconnection encountered can be attributed to the unwillingness of the governments to share their reserves with others, to fears that the existing hydroelectric sources would be exhausted if shared, and to the reluctance to depend on a neighboring country for supplies of energy.[35] For these reasons no regional approach appeared in this sector.

These possibilities of connecting national systems were discussed in a subcommittee on electrification created in 1959 under ECLA. National delegates to the subcommittee also analyzed regional coordination of tariffs and norms and requested technical assistance from the UN in drawing up an inventory of hydraulic resources and for the construction of a regional network of hydrometric and hydrometeorological stations.

Telecommunications

The lack of efficient means of communication among the five countries was acknowledged as one of the major hindrances to the development of the program of economic integration in the area. The first studies in this field were undertaken with the assistance of the UN Special Fund and the World Bank, who hired a French group to study the possibility of building a regional telecommunications network. The study was completed in 1964, and it proposed the formation of a regional corporation to construct, finance, and operate the network at a cost of approximately $10 million. When the national telecommunications authorities met to discuss the implementation of the proposal, they were unable to agree on where the projected enterprise would be located, how its benefits would be distributed, and the relations of the corporation to national authorities and regional institutions. This failure can be compared to the success of COCESNA as an autonomous regional corporation handling a less controversial sector involving smaller funds.

In January 1966, the Ministers of Economic Affairs in the Committee of Economic Cooperation decided that the deadlock must be overcome. In April of the same year, a treaty was drawn up creating a Regional Telecommunications Commission (COMTELCA) to supervise the construction and administration of the network which would be developed by each national authority in its own territory.[36] Thus, the only way in which the deadlock was surmounted was to let each government control and be responsible for that portion of the network constructed in its territory.

The next step consisted in asking the International Telecommunications Union (ITU) for technical assistance to bring up to date the study made by the French group two years before. In July 1967 this was accomplished and the Central American Bank was appointed financial agent for its realization by the national telecommunications authorities. When completed the network was to be 1,300 kilometers long, comprising 960 microwave channels with 33 terminal and repeater stations. Connections with international circuits were envisaged through México and Panama, although the possibility of an independent circuit operating by means of satellites was also considered. It has been projected that the first stage will be completed in 1971 at a cost of $12 million. The Central American Bank intended to use $4 million from the Inter-American Development Bank (IDB) received for this purpose, a supplier's credit covering at least two-thirds of the cost, and, if necessary, its own resources to finance the construction of the network.[37]

The Central American countries supported the creation of COMTELCA because they were able to keep control at the national level and because outside sources provided the financing. Even so, they were warned that "before the countries can proceed effectively with the project on a coordinated basis, they will need to ensure that the coordinating committee has sufficient powers to carry out construction smoothly; it would also have to plan for the next stages of expansion."[38]

Agriculture

Despite the importance of agriculture in the members' economies, the program of economic integration did not affect it substantially. The program's emphasis on industrialization, the rather passive role played by the governments in the agricultural sector, and the fact that the solution to its problems required considerable attacks on traditional vested interests that neither the governments nor the regional institutions were willing to undertake, were the main reasons why agriculture had not been tackled with a vigor commensurate to its importance to economic integration.[39]

The program affected agriculture by establishing free trade for certain agricultural products and by seeking to establish price stabilization in the marketing of basic cereals. Other less important results of regional efforts were related to the regulation of supply of powdered milk, the drawing up of a genealogical register of cattle, the creation of a permanent commission for agricultural extension and research, and some studies related to fisheries development and land tenure.

Joint activities in the agricultural sector date from 1959 when the subcommittee on agricultural economic development was created under ECLA's auspices. But at its only meeting, it did not arrive at concrete results.[40]

As mentioned in the section on free trade, not all the products were immediately liberalized. It is interesting to note that most of the exceptions to free trade were made up of agricultural products. This partially explains the less spectacular growth of trade in raw materials and foodstuffs, which in 1965 represented less than one-fourth of intraregional trade.

Some of the exceptions to free trade were temporary, until certain requisites could be met such as agreement on common tariffs or the signing of specific treaties or protocols regulating supply; this last was the case of basic cereals. The other group of exceptions was made up by the member countries' major exports such as nonpedigree cattle, coffee, and raw cotton, which were definitively subject to the payment of import duties and in some cases to quantitative restrictions.

To fulfill the requirement of the General Treaty for the liberalization of basic grains, the Ministers of Agriculture and Economic Affairs met in Puerto Limón, Costa Rica, in October 1965, to draw up a protocol regulating the supply of basic grains in the region. In almost all the countries, the prices of rice, beans, maize and sorghum were supported by the governments; consequently, the coordination of national price support and agreement on import policies were needed before free trade could be allowed. The protocol dealt with these questions and created a coordinating commission of marketing and price stabilization as a permanent institution to exchange price support information and to control imports in accordance to the surpluses available in the member countries. It also was to make sure that grain supplies imported from outside the area should pay import duties to the national importing agencies and these imports could only take place after consulting the other members on the availability of surpluses. Finally, the permanent commission was charged with the development of a regional program of storage facilities.[41]

With the adoption of these rules when the protocol was ratified, free trade was established for basic grains in the area in February 1968. Although it was too early to judge the operations of the commission, it could be observed that it was mainly charged with the exchange of information on prices and surpluses among the participants and had no financial means to implement a regional policy of price stabilization.

In the case of the regional storage program, the perspectives were relatively better given the possibility of obtaining the necessary financing from the Central American Bank.

Regional observers have concluded that—in the agricultural sector—there did not exist truly regional institutions, but rather discussion forums in which national officials debated regional problems and made recommendations to their governments.[42]

An analysis of the five countries' development plans has revealed that

agricultural programming constitutes the weakest part of the national development plans. A clear definition of development policies needed in this sector is not included. And even less, there is no attempt to delineate and encourage the productive specialization that the process of regional integration demands.[43]

Apart from these limited results, the member countries did agree to establish a coordinated policy to control imports of powdered milk, by which quantitative restrictions were to be imposed by the Executive Council of the General Treaty. But when the moment arrived to determine the amounts that each country could import, no agreement could be obtained from the national delegates. Other activities in this sector consisted of the preparation of studies, although "the concrete results obtained... [bore] no relation to the extent of the studies nor with the number and scope of the recommendations that have been approved in the meetings."[44]

This situation worried regional officials, and foreign assistance was being sought to surmount it. In the field of price stabilization and marketing of basic grains, the members intended to create a fund of $25.4 million of which one-half would be income from sales of wheat obtained from the World Food Program and the other half members' contributions.[45]

Monetary Cooperation

Discussions on the joint treatment of monetary questions started almost at the same time that ECLA's efforts began in the region, but took place independently and at a different level. Some cooperation among central banks occurred when Guatemala's central bank, dating from 1946, collaborated in the creation of the central bank of Honduras in 1951. Under Honduras's initiative, the central bankers from the five countries met in Tegucigalpa for the first time one week before the Ministers of Economic Affairs met in the same place in August 1952. This first meeting was of an exploratory character since the central bankers met under the condition that "they would not adopt resolutions that would compromise the internal economic policies of the participant countries."[46]

From then on they met several times and tried unsuccessfully to establish a regional payments system. However, it was not until the increase of intraregional trade, the signature of the General Treaty, and the creation of the Central American Bank that the central banks decided, at their sixth meeting held in Tegucigalpa in July 1961, to create a regional clearing house.

The clearing house was conceived originally as a financial institution rather than one restricted to the payments arising from intraregional trade. The participants agreed to contribute to a guarantee fund and to a fund of current operations. The first required the deposit of $75,000 from each member to guarantee the convertibility of balances engendered by intraregional trade; the other was formed by the equivalent of $225,000 by each member in its own currency, to finance intraregional trade in local currencies.[47] In other words, "it was as if the Central American countries had formed an international institution that would grant and receive credits with the initial contributions and the various registered operations."[48]

After less than two years, in 1963, the clearing house agreement was modified because the guarantee fund was never used, "the creditor and the debtor banks had mutually agreed to settle the dollar balances among themselves, without resorting to the Fund."[49] The members' contributions were abolished and transformed into credit lines of $500,000, with the compromise that balances beyond this limit would be settled in dollars twice a year and that any excess beyond the credit lines before would have to be settled weekly in dollars also. Each central bank was authorized to extend credit unilaterally beyond the limit at 3.5 percent interest.[50]

These changes constituted a step backwards in the path of monetary cooperation among the members, because they transformed the regional institution into an accounting office for the recording of balances among countries. Even so, the clearing house increased the use of local currencies to finance intraregional transactions; it compensated 48.8 percent on intraregional visible trade in 1962 and 84.5 percent in 1967. Thus, its activities represented economies in the foreign exchange reserves of the participants and promoted contacts among the central bankers that led to the coordination of certain unilateral measures adopted by the governments to protect their balances of payments, as in the case of the imposition of exchange controls by El Salvador and Guatemala in 1962.[51]

But these achievements in the field of multilateral payments were considered insufficient when compared to other more ambitious steps taken in other sectors. The need for stronger measures was frequently discussed among the central bankers and led to the decision, made by the Central American Presidents during the meeting with President Kennedy in March 1963, to establish a monetary union. In February 1964, the presidents of the central banks signed an agreement "to promote the coordination and harmonization of monetary, exchange and credit policies, so as to create gradually the basis for a Central American Monetary Union." The agreement also institutionalized the meetings of the central bankers which had taken place regularly in the past, and created an executive secretariat.[52]

The lack of coordination of the agreement with other regional activities provoked certain criticisms at the time. González del Valle has explained this absence of coordination as follows:

On the one hand, the national banking systems have traditionally enjoyed a high degree of political and administrative independence in Central America, where the General Treaty contemplates the adoption of basic policy decisions at the ministerial level. On the other, the essentially "structural" (as opposed to "financial") approach to economic integration originated from the notion that trade barriers rather than exchange and monetary obstacles had to be removed before a common economic policy with meaningful results in the industrial, foreign trade and communication areas could be designed.[53]

To these two reasons could be added the members' relatively favorable balances of payments at the time of adoption of the agreement, due to an improvement in coffee prices and the appearance of cotton as a major export for some of them. Nevertheless, with the decline of export prices in 1965 balance of payments cooperation began to appear as one of the main functions of the Monetary Council, and consequently, the first steps to coordinate its activities with other regional efforts were initiated. At a joint meeting of Ministers of Finance and Economic Affairs, the Secretariat of the Monetary Council affirmed that "a very marked tendency toward disequilibrium in the external sector of the Central American economies is observed, which makes urgent the adoption of decisions to strengthen it, as an indispensable condition for the development of the integration program in the Isthmus."[54]

In 1966, Costa Rica imposed differential exchange rates to control imports and did not except intraregional trade from these measures. The Monetary and the Economic Councils urgently met in Tegucigalpa in January 1967 to discuss this first balance of payments crisis that affected the program of economic integration. The Costa Rican delegation declared, among other reasons, that "the lack of flexibility in the use of certain other instruments as a normal consequence of the process of Central American economic integration"[55] had forced them to take this step. The problem was solved when Costa Rica agreed to except intraregional trade from the emergency measures it had imposed.

This first reaction to the unilateral action taken by one country to defend its balance of payments marked the beginning of the regional treatment of one of the most acute problems facing all the participants. The next step consisted of a study by the Permanent Secretariat of the General Treaty (SIECA) and the Executive Secretariat of the Monetary Council on the balance of payments situation of the region. This was discussed at a tripartite meeting, which gathered for the first time the Ministers of Finance and the Monetary and the Economic Councils. Very few concrete results, however, were derived from it.[56]

In June 1968, a second tripartite meeting signed a "protocol" to the General Treaty, creating a surcharge of 30 percent on the import duties of a list of products and authorizing the members to impose consumption or sales taxes on another list.[57] Although Costa Rica refused to ratify the protocol—because its balance of payments situation had improved in the meantime (which caused a break in the common external tariff)—these measures represented important steps toward monetary and fiscal cooperation in the area. Regretfully, though, they constituted the only instance in which the participants adopted joint actions in this sector. Meanwhile, because of the lack of specificity in the terms of reference of the Central American Monetary Council, other methods of adjusting the balance of payments remained within the hands of the zealously independent central bankers.

Finally, a project to create a stabilization fund, as one of the short-term objectives of the Executive Secretariat of the Monetary Council, was endorsed in several regional meetings. It is of interest because it depended on the possibility of obtaining external assistance to finance its operations. The fund was to be made up of participants' contributions—the Special Drawing Rights of the International Monetary Fund (IMF) and support of the United States government—and would allow a partial pooling of foreign exchange reserves in the area. The Ministers of Economic Affairs stated that

the degree to which the stabilization fund achieves its objectives will depend in a great measure on the amount of external assistance that will be obtained for such a purpose, since the internal contributions of the Central Banks will not be sufficient, even though they represent a relatively great effort on the part of the Central American countries.[58]

Some opposition had begun to be heard against this proposal, which remained to be approved.

Regional Financing

Although special consideration has been given in this chapter to the financing of specific projects, regional financing is here treated separately since the member countries have created a regional institution dedicated exclusively to the financing of projects of regional significance. Thus, this section will deal with the Central American Bank, other sources of official foreign financing for regional activities, and private foreign investment.

The origins of the Central American Bank were discussed in Chapter 3. Here I will only mention that the agreement creating the Bank was signed by four countries—Costa Rica joined in December 1963—"to promote the economic integration and balanced growth of the member countries." Its initial authorized capital was $16 million, which was raised to $20 million with Costa Rica's participation. By April 1969 increases had brought it to $60 million, of which the members had effectively paid $20 million. A protocol creating a guarantee capital of $40 million was signed to back the first operations of the Bank in the world's capital markets.

But the Bank's most important achievements were not in the mobilization of regional resources, which were in any case scarce. Rather, it can be better appreciated for its ability to attract foreign assistance. As of April 1969 the Bank's overall resources amounted to $250 million, of which $215 million (or 86 percent) came from foreign sources—about three-fourths from the United States and the Inter-American Development Bank (IDB), where the United States has decisive influence, and the other fourth from suppliers' credits granted by some Western European countries and México. As of the same date, the Bank had authorized loans amounting to $150 million, $88 million (or 59 percent) for the construction of the regional "infrastructure", $52 million (or 34 percent) for industry, and $10 million (or 7 percent) for housing. It has been estimated that one-fourth of the loans were financed from the members' contributions and the rest from foreign sources, but the situation changes by sectors, especially in housing where all the funds came from a US-AID loan of $10 million.[59]

One of the main problems that the Bank faced was the absence of a regional investment policy. Due to the impossibility of establishing regional priorities among the members, the Bank independently established its own guidelines without coordinating them with other regional efforts. Nevertheless, the sectors influenced by the Bank's credit activities are in accordance with the program's main goals, with industrialization and the construction of a regional "infrastructure"—roads and telecommunications—being the most favored.

The next most important source of funds dedicated exclusively to regional activities was the UN, following the priorities outlined by ECLA and the specialized agencies. From 1950 to 1966 the financial assistance granted by the UN to the program of economic integration amounted to $22.4 million to support regional efforts in transport, electricity, education, telecommunications, agriculture, fisheries, and the functioning of certain permanent regional institutions. The individual contributions of the member countries to match the UN regional assistance amounted to $5,000 per annum from 1953 to 1965 and $6,000 per annum from 1966 on, for a total contribution from 1953 to 1966 of $335,000, or less than 2 percent of the total amount received from the UN.[60]

In addition to these flows of official foreign assistance in support of the regional program, private foreign investment also increased. Although this increase cannot be fully attributed to the regional program, it is assumed that integration played a certain role in attracting such funds. Figures in this field are somewhat elusive; it has been estimated that private long-term foreign investment came into the area at a rate of $17 million a year during 1955-1959, and that this rate increased to $24 million for the period 1960-1963. The declared value of direct private investment from the United States in the region—by far the largest supplier—increased from $340 million in 1958 to $375 million in 1963. This investment took place traditionally in agriculture and public utilities, but some changes in its distribution were observed, with manufacturing and oil refining attracting the largest proportions—$28 million during 1962-1965.[61]

The conditions under which these increases of private long-term foreign investment arrived in the region were not regularized, since no regional policy for their treatment emerged within the program of economic integration. On the contrary, the members competed by offering tariff protection and tax holidays to attract the largest amounts of these funds. This was especially true in the manufacturing sector, where traditional exporters, trying to jump the regional tariff barrier, established industries that performed a very small proportion of the value added to their products within the area to benefit from the expanded market. The acquisition by large foreign firms of existing industries owned and administered by Central Americans, raised doubts about the fulfillment of one of the main justifications for direct private foreign investment—that it brings new technology and know-how into the area. None of these questions was given regional consideration; meanwhile the competition among the member governments persisted.[62]

To conclude, while the foreign lending agencies authorized loans to the individual countries amounting to $531.2 million during 1961-1967, they supported the regional program with loans amounting to $185.5 million from 1961 to June 1968. The UN furnished technical assistance bilaterally to the five countries that amounted to $43.3 million from 1950 to 1966, while it supported regional activities with $22.4 million in the same period.[63]

These figures show that despite the vital role that external assistance played in the evolution of the program of economic integration in Central America, bilateral aid was far more substantial, and that the participants continued placing their national concerns above their regional efforts.

It is apparent that in the near future this situation will persist given the unpromising outlook for the region's major exports in the world market. This situation has been summarized as follows:

it appears feasible for the countries of Central America to sustain a rise in public investment in the face of lower rates of export growth in the next few years, provided foreign financial assistance for project loans is forthcoming and is matched by adequate internal saving efforts.[64]

Planning

ECLA insisted from the beginning that regional industrialization be based on a mutually agreed plan so as to avoid investment duplication and to permit the balanced growth of the members. Because of the absence of planning at a national level, it was some time before planning could start at a regional level. Although ECLA and the World Bank had worked out some suggestions to almost all the participants individually, the necessity for national plans came in 1961 with the Alliance for Progress, which made planning a requisite for obtaining its assistance. As a result the Central American governments started to abandon their traditional laissez faire conceptions of their role in the economy and began to envisage establishing priorities in public investment to secure the assistance offered by the Alliance. Also that part of the General Treaty which dealt with external financing raised the question of a regional approach in planning. It is significant that one of the first resolutions adopted by the Economic Council at the suggestion of the institutions charged with the administration of the Alliance, requested that the Organization of American States (OAS), the Inter-American Development Bank (IDB), and ECLA create a regional mission to furnish technical assistance to the Central American governments in drawing their national plans and coordinating them at a regional level.[65]

The Joint Planning Mission for Central America was thus formed with experts from the OAS, the IDB, and ECLA, and started operating in 1963 under the supervision of an advisory committee formed by representatives from these three institutions, the Secretary General of the integration program, and the President of the Central American Bank. Two-thirds of its budget was financed by OAS-IDB-ECLA, and the other third by SIECA and the Central American Bank.[66]

Since the most urgent need at the time was the drawing up of national plans, the first task of the Mission was to provide technical assistance to the governments in the organization or to create national planning offices. At the same time, the Mission proposed a uniform methodology for the preparation of the plans; they were completed in 1965 and immediately submitted to the Committee of Nine of the Alliance for Progress.

That the Planning Mission was created to obtain foreign assistance is evident. Even so, the finished national plans revealed the inexperience of the individual countries in this field, and also once again the impossibility of establishing regional priorities in those sectors entrusted to common efforts. The conclusion of the Committee of Nine was that the Mission's activities represented only "the beginning of a process of regional planning."[67]

Technology

In the belief that joint efforts in the field of technology would reduce costs and improve efficiency, the Central American Institute of Industrial Technology and

Research (ICAITI) was created in 1955 with the support of the UN. Its functions were restricted to industry, having acted as a consultant on questions such as the application of the Regime of Integration Industries, the setting of quality norms, and the establishment of the origin of free trade products; it also advised private entrepreneurs. ICAITI merely acted on requests from regional institutions and private entrepreneurs; it did not establish a set of goals for this vast sector.

As with other joint ventures, ICAITI's main problem was financial; the member countries did not fulfill their pledges to match the considerable support that it received from the UN. This situation provoked an almost perpetual financial crisis in the institution which was reflected in the piecemeal approach that characterized its activities.[68] Thus, despite the urgent needs of the five member countries in this field, their joint efforts had limited results, again due to their reluctance to finance regional activities. ICAITI constitutes an example of the limitations of joint efforts inadequately supported by the member governments.

Public Administration

In 1956, at ECLA's suggestion, the Central American Ministers of Economic Affairs set up a school to strengthen the national bureaucracies of each country so they could meet the demands of the program of economic integration. The Central American Institute of Public Administration (ICAP), before known as the Central American School of Public Administration (ESAPAC), operated initially at an academic level, but in 1966 it was decided by the Ministers of Economic Affairs that the courses should be more directly relevant to the needs of the program of economic integration, in such fields as trade, customs, transport, electricity, telecommunications, and so on. ICAP was supported by the members' contributions, the UN, and AID-ROCAP. Like ICAITI it had problems with members' contributions; even so, by 1966, 1400 government officials had taken courses at the institute.[69]

Other Sectors

Along with the efforts toward economic integration in the area, attempts to integrate other fields have appeared, such as nutrition, where a research institute was created; university education, where a permanent secretariat and regular meetings of the rectors of the national universities were set up; and within the Organization of Central American States (ODECA) where some concrete questions were handled regionally such as labor conditions, the unification of curriculums in primary and secondary education, a regional textbook program, and so forth. These efforts were made outside the strictly "economic" approach that inspired the Central American technocrats. Their relationship to the "economic" institutions will be discussed in the following chapter.

Having surveyed the main sectors which were affected by the process of regional integration, it is necessary to attempt an evaluation of its impact (during the period under analysis here) on the overall process of economic development of the participant countries.

This can be done by observing not only what integration affected directly, but also those activities that have been left out of the integrative effort, with special attention to the relative weight of all these sectors—those left out as well as those affected—in the process of economic development of the region.

To begin with what was excluded from integrative activities the absence of concrete achievements in the agricultural sector of the countries' economies appeared as the main limitation to the relevance of integration. This limitation is more evident if it is recalled that agriculture employed almost two-thirds of the region's economically active population. Consequently, any policy that intended to encourage the participants' industrialization through the expansion of their national markets had to confront the question of consumption capacity of the largest sector of their populations: the Central American peasants. In effect, even the expanded market was small considering its purchasing power. It represented 15 million people, increasing at a rate of 3.5 percent annually, with a regional average per capita income of $305 annually. The situation was even worst when it is considered that the subsistence sector of Central American agriculture had an average regional per capita income of between $70 and $100 annually.

The absence of any concrete results from the integration process in the agricultural sector of the region's economy was the main limitation to its relevance for the solution of the most urgent problems of the area. The meager results accomplished in the field of free trade in basic grains illustrate this situation. The absence of any effort within the regional process to enforce an agrarian reform program, which constitutes an essential requisite of any policy tending to increase the level of income of the Central American population, restricted the process' impact.

Nevertheless, it could be argued against the preceding assertions that economic integration was not a panacea that would solve all the problems existing in the region, and that those problems affecting the agricultural sector of Central America should be the concern of the individual governments. Furthermore, it could be added that the process of economic integration should be judged by what it had done and not for what it had left out.

An evaluation of this sort demands a considerable degree of abstraction since it means to set aside more relevant questions—such as those related to agriculture. It allows the conclusion that the process of regional integration did not contribute to the solution of the most urgent problems affecting the majority of the Central American population, and also it allows the conclusion that those achievements which could be credited to the integration process had a limited impact, restricted to the urbanized sectors of the region.

What was the impact of the process on those sectors that it had affected?

The main field of integrative activities was the members' industrialization by means of the expansion of intraregional trade. If the figures of the increases in intraregional trade are used as the measure of integration's success (as it was frequently done in the region and abroad), it is easy to conclude that the process was successful. But this again left aside the more important question of the kind of industrialization that was provoked by the increases in intraregional trade. Trade in Central America, as mentioned earlier, was not being encouraged for trade's sake; it was considered an incentive for the members' industrialization. Thus, the real impact of the Central American integration process has to be evaluated in the light of the sort of industrialization that it provoked.

There are several reasons to conclude, from this point of view, that the process in this sector also had limited and, in some cases, negative effects. It was limited because of the relative weight of the manufacturing sector in the Central American economy, which employed between 10 and 15 percent of the region's economically active population. But it was still more limited because, as it was mentioned earlier, less than 20 percent of the countries' industrial output was traded regionally. And it was negative because of the disorderly way in which the growth of the manufacturing sector had taken place under the program of economic integration. Thus, instead of a regional policy of industrial development, the governments competed among themselves to attract the largest proportions of investment into their countries. This allowed the establishment of "make-up" industries which performed a very low proportion of the value-added to their products within the region, and which were owned by traditional exporters who tried to jump the tariff barrier established by the integration program. Also, the absence of a unified policy of industrial development favored mostly foreign investors who came to the region looking for the best deal. Finally, besides the decrease in fiscal revenues that this policy—if it can be called so—caused, and the strain that it imposed on the members' balances of payments, the differences among the participant countries were also aggravated, since most of the industrial investment went to the relatively more developed: Costa Rica, El Salvador, and Guatemala.

From the preceding considerations, it can be concluded that the impact of the process of economic integration on the development of the participant countries should not be exaggerated. This is particularly evident when those sectors excluded from integrative activities are observed. And those sectors that were affected by the integration process show that its impact was limited in the urbanized sector of the region, and that it was also negative, as shown by the sort of industrialization it provoked.

Thus, the process of Central American economic integration did not contribute decisively to change the economic and social profile of the region, since it was irrelevant to the solution of Central America's most urgent problems.

5 The Institutional Setting

If numbers of institutions were an index of integration, Central America would rate first, because its integrative efforts have provoked the establishment of many different sorts of multilateral and permanent organizations.

This chapter contains an analysis of the institutional setting in which the integration program took place and complements some existing studies.[1] I will examine the evolution of regional institutions and focus on institutional autonomy as an indicator of the occurrence of "the process of transferring exclusive expectations from the nation-state to some larger entity."[2]

Institutional Evolution

As noted in Chapter 2, the first stage of the Central American process was characterized by what we called ECLA's hegemony, consisting of the ideological justification and the material support that the regional commission granted to the process. The institutions created during the period 1951-1959 were: (1) the annual meeting of Ministers of Economic Affairs; (2) periodic subcommittee meetings, where national delegates met to discuss those questions that required technical elaboration before being submitted to the Ministers, such as tariff equalization, lists of products to be granted free trade, transportation, electric power, and so forth; and (3) two permanent institutions—ICAITI, created to deal with industrial technology, and ESAPAC (known today as ICAP), charged with the training of national bureaucrats. ICAITI and ESAPAC were financed in part with funds from the United Nations. Each had its own board of directors, either made up of representatives of the Ministers of Economic Affairs—as in the case of ESAPAC—or the Ministers themselves—as in ICAITI. The Executive Directors of both institutions submitted annual reports to the Committee of Economic Cooperation, which made for a certain degree of coordination with the rest of the program. In addition there was the control exercised by the UN and its agencies through financing.

At this stage the program "spilled-over" toward sectors which were complementary to import substitution. Thus, the Multilateral Free Trade Treaty of 1958 was accompanied by the signing of the treaty establishing the Regime of Integration Industries and was followed in 1959 by the tariff equalization treaty to protect mainly those industries which, it was hoped, would be established regionally. ECLA's México Office enjoyed considerable freedom of action be-

67

cause its functioning was not financed by the member governments but by the UN. Although it had the monopoly of the initiative, backed as it was by ECLA's prestige and technical skill, it always secured governmental approval of its proposals. This is illustrated by the fact that the proposals and the technical assistance needed to implement them were always unanimously approved by the Committee of Economic Cooperation and later ratified by each government.

At that time, the program appeared to be a cooperative effort to establish a number of industries enjoying preferential treatment and protected by unified tariffs. The possibilities for increased politicization were scarce because of the passive role played by the local technocrats. But actually, the Secretariat carefully avoided any activity of this sort. By the end of the period when more permanent institutions were about to be created, it was generally agreed that the question of more comprehensive levels of integration—to say nothing of political union—should be left to the very distant future and should be conditioned by the fulfillment of the initial goals and the equitable distribution of the benefits among the participants. There was the general feeling that "economic development" was more important than anything else.

No thought was being given to the creation of an international secretariat invested with executive functions,[3] which was considered necessary to the process of increased politicization. Instead, two types of institutions that the treaties of 1958 and 1959 created—industrial and trade commissions—were multilateral organizations where governmental representatives would sit, and it was to them that ECLA's México Office planned to transfer its activities "at a product level," to be able to devote itself to questions of more "general scope." This gradual transfer of tasks was in line with the general UN policy of building self-sustaining projects that could be entrusted to local authorities. But the possibility of increased politicization under these conditions was almost impossible and was intentionally set aside.

These new institutions never saw the light of the day since the process underwent radical changes before the treaties creating them were ratified. Instead, under the General Treaty, permanent local institutions were created. The new structure was, to a certain point, a duplication of the old one; an Economic Council was created, formed by the Ministers of Economic Affairs, "to direct the integration of the Central American economies and to coordinate the economic policy of the member states." This organization was also charged with "facilitating the execution of the resolutions of the Committee of Economic Cooperation" (Article XX of the General Treaty). This made it possible for the new institutions to carry out previous directives with the advantage that no hierarchical problems would be raised, since the Ministers of Economic Affairs participated in both and kept open ECLA's channel of communication with the governments. An Executive Council was also created, where officials at a lower level—the "Vice-Ministers" of Economic Affairs—would meet to prepare the decisions for the Economic Council and to apply and administer the General

Treaty by determining the measures needed to fulfill its provisions. It also assumed the tasks of the industrial and trade commissions (Articles XXI and XXII of the General Treaty).

Both the Economic and the Executive Councils could still be viewed as part of ECLA's plan for a gradual transfer of tasks to local institutions. The Permanent Secretariat (SIECA), created to oversee the application of all integration treaties signed by the participants, as well as the fulfillment of the resolutions adopted by the councils and the realization of the tasks set by them (Article XXIV of the General Treaty), raised certain problems of duplication since some of these tasks had been previously performed by ECLA's México Office. However, their coordination was assured by ECLA's participation in drafting the General Treaty, which explains the rule previously mentioned, that the Economic Council would facilitate the execution of the resolutions of the Committee of Economic Cooperation. In addition the first two Secretary Generals were ex-officials of ECLA's México Office. These conditions led to an agreement between ECLA and SIECA in which ECLA's México Office "would be occupied, from then on, with the preparation of those studies required for the program's long-term orientation, while SIECA would give priority to short-term tasks."[4]

Finally, the Central American Bank of Economic Integration was created "as an instrument for the financing and promotion of integrated economic growth on the basis of regional equilibrium" (Article XVIII of the General Treaty). But a separate treaty was signed outlining the functions and operations of the Bank, without mentioning the coordination of its activities with other regional institutions, even though the Ministers of Economic Affairs and the presidents of the central banks were members of its board of governors.

It is important to examine the consequences of these changes which took effect at the ratification of the General Treaty and which provided the institutional setting for the program's next stage.

One of the first consequences was a diminution of ECLA's activities in the region, illustrated by the fact that the annual meeting of the Committee of Economic Cooperation only took place twice after 1962—in January 1963 and January 1966. And the meetings of the subcommittees, besides concentrating on specific issues, occurred less frequently than during the period of ECLA's hegemony. But probably of more importance was the decrease in ECLA's México Office short-term activities. And the long-term tasks of the regional commission were limited to studying the possibilities for establishing a customs union, a common commercial policy vis-à-vis third countries, industrial and agricultural complementarity, the construction of the regional "infrastructure," and the implications of these measures for the formation of a regional economic union linked to neighboring countries and to the projected Latin American common market. Thus, as planned, ECLA assumed a more analytical and consultative role on long-term questions, whose implementation was left to local institutions. But this also meant a considerable reduction in the human and material resources devoted to Central America, since the subregional office increased its attention to México and the Caribbean countries at the expense of the Central American integration process which, in the previous decade, had absorbed almost all its resources.[5]

The other consequence of the events of 1960 on the institutional domain, therefore, should be sought in the functioning of the new institutions, particularly the Economic and Executive Councils and the Permanent Secretariat (SIECA). With their creation, the possibility for politicization increased. This was particularly true of SIECA because it was made up of a nucleus of local technocrats with executive authority. However, because its activities were limited to implementing decisions made in the Councils, this possibility was hindered. In fact, day-to-day problems drowned the Secretariat in a mass of details so that its principal task became "the putting out of small fires." As long term projections were made by ECLA's México Office, there was no need to rise above these "small fires." In addition the direct submission of the new institutions to governmental control (not true of the old structure), because government contributions were one of their main sources of funds, only aggravated this situation and forced a cautious approach to propositions that could diminish the role of the governments in the program. Thus, politicization was hampered by the increase of governmental control which tacitly imposed on the regional technocrats the need to avoid those proposals that could mean a decrease of these controls. This situation, added to their immersion in day-to-day problem solving, made the technocrats' main goal the preservation of the program and consequently, themselves. At this point they had no interest in political action.

But this is not all; there was also the absence of a global approach. This is shown by the haphazard creation of other institutions charged with the regional treatment of particular economic sectors. This was mainly due to the uncoordinated efforts of governmental agencies, which gave the impression that each was trying to "integrate" its own sector of activities while paying little attention to the existence of other institutions with which it frequently overlapped or collided.

The practice of setting up institutions by economic sectors originated in the Committee of Economic Cooperation under ECLA, when ICAITI and ESAPAC were organized. But the intensification of the program provoked a proliferation of uncoordinated sectoral institutions. One of the most important examples of this disorderly development of joint efforts is offered by the Central American Bank of Economic Integration. It was created principally to obtain external financial assistance. It was only at its creation that its financial activities were linked to the rest of the program because, to obtain its funds, each government was required to ratify the regional treaties previously signed (Article XXXIV of the Constitutive Agreement). This assumed that subsequent activities could be left without the constant coordination that the Bank's financing demanded. And although the sectors to which the Bank was devoted were drawn rather widely in the Constitutive Agreement so as to enable the Bank to exist coherently within the program as a whole, its functioning revealed that they were carefully separated from the rest.[6]

Another case in point was the isolated existence of the Monetary Council where the presidents of the central banks met to work out the plans for a monetary union in the region. And even when their isolation could be explained by the independence which the central banks enjoyed in their own countries, the fact is that, regionally, their isolation was only repeated. The only case of cooperation between the Monetary Council and SIECA was during the balance of payments crisis, experienced by almost all the countries with the decline in export prices that began in 1965 and that led to a protocol adopting emergency measures. But this was their sole joint action, which leads to the conclusion that other joint efforts depended on crises or "small fires."

Other examples are the increased isolation of ICAITI and ICAP. The Central American Corporation of Air Navigation Services (COCESNA) did ask SIECA to recognize it as part of the program of "economic" integration, which meant that SIECA would have the responsibility of overseeing its activities and achievements. But SIECA refused on the grounds that this was the function of COCESNA's president.[7]

Nevertheless, there are some cases in which institutions were made part of SIECA, such as the transformation of the Joint Planning Mission into its Development Division; or when SIECA assumed the secretariat of periodical meetings, as those of the national price stabilization agencies; or when meetings were organized under its aegis as those of the planning and customs directors (these were mainly meetings where governmental officials exchanged information on their national activities, while the most important regional permanent institutions continued to function independently, as illustrated by the projections of three different institutions on regional road transport); or when the studies were done by several institutions on the concrete possibilities for industrial development; and when the studies were undertaken by the Central American Bank on institutional financing.

In contrast the very few cases of interinstitutional cooperation were characterized by their short-lived existence, since they consisted of meetings where precise questions were discussed without arriving at decisions to be implemented or guidelines to be coordinated. And after they had taken place, the participant institutions turned back to their independent efforts. This was the case when the three institutions drew up the road network plan without agreeing on coordinated measures to survey its construction.[8] Another case was that—in spite of the industrial projections introduced by the Joint Planning Mission and discussed in two meetings with representatives from the Central American Bank, SIECA, the OAS, ICAITI, ECLA—each participant continued to make its own projections, although only those of the Bank were implemented.[9]

Some regional officials have argued that coordination existed among the activities of what were considered "economic" institutions, because the Ministers of Economic Affairs were intergovernmental directors of the Central American Bank, IACITI, and ICAP, and sat in the Economic and Executive councils, and also because the Bank's Assembly of Governors was made up of the presidents of the central banks and the Ministers of Economic Affairs which allowed a coordination of financial and monetary questions. Furthermore, it was pointed out that the participation of delegates of the permanent institutions as observers in almost all regional meetings assured a certain degree of coordination. But this was all at a very high and general level and did not take into account the day-to-day activities, where duplication appeared and where the few initiatives the technocrats could take were weakened by their fragmentation. Rather, the initiative was left to the supreme authorities, while a sort of "sectoral sovereignity" appeared among the permanent institutions.

If we look away from those efforts considered "economic," ODECA's weak but separate existence shows that not even the initial separation between lawyers and economists was surmounted; the former continued to try to control the program of economic integration, and the latter continued to carefully avoid these controls. ODECA's failure, compared to the relative success of economic integration, justified the persistence of this situation but at the same time represented an obstacle to task expansion of the "economic" program, particularly when it meant that the Ministers of Foreign Affairs had to lose control of certain activities that were traditionally within their jurisdiction. These possibilities were naturally resisted, not only because the Ministers of Foreign Affairs felt that they should direct all integrative efforts, but also because giving away some of their traditional diplomatic activities diminished their relative weight within each national administration, already considerably affected by the control exercised by the development agencies on the "new economic diplomacy" of which economic integration was an important part. These activities, in the technocrats' opinion, required more skill than formality.

The establishment of a common commercial policy as a subsequent step to the setting up of the common external tariff illustrates this situation. Although it cannot be said that a complete common tariff existed, nor that a common commercial policy was established, some initial steps were taken such as: the joint participation of the five countries in the two UNCTAD meetings and in their preparatory sessions; the negotiations for the possible participation of Panama and the formation of a Latin American Common Market at which SIECA participated, representing the five Central American countries; and the collaboration of SIECA and the Central American Bank in promoting exports. These steps were all based on decisions of the Ministers of Economic Affairs, who stated that "the formulation and superior direction of the common commercial policy of the member countries of the General Treaty of Central American Economic Integration toward the rest of the world is the responsibility of the Economic Council," and that "the Executive Council is charged with applying and administering the common commercial policy formulated by the Economic Council."[10] The Ministers of Foreign Affairs on their side approved a series of resolutions asserting what role they felt they should play in the formulation and execution of the common commercial policy.[11] This concrete example confirms the persistence of the cleavage among both governmental agencies. In the meantime lawyers and economists were lost in a quarrel, at times ridiculous, which hindered the possibilities of coordinating their efforts toward a common goal.

Furthermore, ODECA has tried to orient its activities toward more "functional" areas, by what has been called a "demonstration effect spill-over,"[12] particularly toward those that, for one reason or another, were left out of the "economic" program, such as the equalization of labor conditions as a requisite for labor mobility. The taking over of this area by ODECA effectively excluded the economic institutions from this field, even when they recognized it as one of their goals.[13]

The evolution of regional institutions illustrates the simultaneous development of a regional consensus and of a "sectoral or ministerial sovereignty" that was to become an obstacle to spill over and task expansion into other sectors. Even so, some still saw in this uncoordinated proliferation of regional institutions and sectors a parallel march toward a goal purposefully left ambiguous, and believed that the passing of time would make them converge to form the larger entity. This wishful projection of the future, found frequently among regional officials, would have been feasible if the permanent institutions had acquired a considerable degree of autonomy from the member governments, an autonomy which would leave the coordinating of regional efforts to the institutions themselves. Consequently, the degree of institutional autonomy that emerged within the process is an indicator of the transfer of expectations from the nation-state to the larger—though still uncoordinated—entity.

Institutional Autonomy

Autonomy will be understood here as "the capacity to act independently of the member governments, and particularly the financial capacity" that the different institutions acquired.[14] The analysis will include the study of three questions: first, the principal sources of institutional financing; second, the degree of participation of different social groups in institutional activities; and third, the pattern of outcome of the decision making process in two specific issues.

Institutional Financing

The budgets for one year of four regional institutions selected for their relative importance will be discussed here.[15] The Bank and the Planning Mission are not included because their financing was looked into in Chapter 4. These four institutions are: the Permanent Secretariat of the General Treaty (SIECA), the Organization of Central American States (ODECA), the Institute of Industrial Technology and Research (ICAITI), and the Institute of Public Administration (ICAP).

In SIECA's budget for 1968, governmental contributions represented 60 percent, followed by external assistance from AID-ROCAP, 38 percent, and the rest (2 percent) mostly derived from the sale of publications. Governmental contributions to SIECA increased considerably after its creation, from $50,000 per member to $120,000, but even so, external assistance played an increasingly important role.

ODECA differs from the Permanent Secretariat (SIECA) in the order of importance of these sources in its financing but resembles it in the absence of resources of its own. ODECA's 1967 budget was financed 60 percent by AID-ROCAP, 30 percent by governmental contributions, and the rest by the Spanish government (8 percent) and the Organization of American States (OAS) (2 percent). The salient characteristic of ODECA is its dependence from external assistance for the financing of more than two-thirds of its budget. It should be noted that another difference with SIECA is that governmental contributions to ODECA have remained constant.

The budget of the Central American Institute of Public Administration (ICAP) in 1966 was 40 percent financed by the UN, 30 percent by AID-ROCAP, and 25 percent by the member governments. The balance came from the sale of publications. It differs from ODECA in the sources of external assistance, which can be understood by remembering that ICAP was created when the UN played an important role in regional integration due to ECLA's involvement.

ICAITI was the only institution to derive its financing equally from its sources. Its 1967 budget was covered one-third by the UN Special Fund under an agreement lasting until June 1968, one-third by the governments, and, uniquely, one-third from its own resources.

The most important conclusion to be gained from the observation of the financing of these four regional institutions is that they did not enjoy financial autonomy, and that the governments were still the main source of income for most of them. In those cases where the governmental contributions were relatively less, instead of generating their own resources, these institutions sought the assistance of foreign sources, mainly the United States government. (ICAITI can be expected from this assertion, given the support of the UN, although the end of the agreement with the Special Fund meant that it was approaching a financial crisis.)

Because of the reluctance of the member governments in effecting their contributions, the question of finding autonomous sources of financing was not the main preoccupation of these institutions, but simply finding ways to assure the institutions' continuous operation. Here external sources proved more reliable than the governments.[16] This evidence supports the general assertion made in preceding chapters of the importance of external factors in initiating the program and also in maintaining the regional institutions.

It might be deduced that the situation is such because the financing of regional institutions was too heavy a load for the individual governments given their problems in financing their own budgets. However, a comparison of the public expenditures and human resources required by the five central governments with regional expenditures and resources, excluding initial contributions to the Central American Bank, dismisses this explanation. In 1966 the contributions of the five member governments to all regional institutions represented 0.4 percent of the public expenditures of the five central governments, and while the governments employed 4,916 professionals—excluding superior and intermediate executives—regional institutions were staffed with 202 professionals—including executives.[17]

The relative demands of the regional activities, viewed from this angle, were almost insignificant, but still the governments failed to make their contributions on time. Perhaps this is an indication of the importance that the governments assigned to the regional institutions.

Participation

Another indicator of institutional autonomy is the degree in which social groups support the institution and transfer their loyalties and address their demands to the larger entity. This support is particularly important to the integrative efforts of developing countries, because theoretically, a larger base would give the technocrats a larger margin of maneuver, since they can arrange different coalitions to advance integration with less governmental interference.[18]

Unfortunately, the program in Central America attracted very limited support. Its main goal—the participants' industrialization—was the principal factor limiting the degree of participation in the program, because of the small importance that industry had in the member countries. The exclusion of agriculture meant, among other things, that the largest, as well as the most powerful sector of Central American societies did not participate in the integration process. Consequently, the dialogue that it provoked was limited—in number as well as in relative weight—to a handful of technocrats, industrialists, and bureaucrats who represented the middle and upper strata of these societies and reflected their particularistic social structures. Thus the transfer of exclusive expectations from the nation-state to the larger entity in the region was hindered by the degree to which the program was unable to transcend or differ from the national social structures of the member countries.

Within this limited context, only those sectors capable of articulating their demands participated, and authorized observers concluded that the participation of individual entrepreneurs has been particularly relevant; some entrepreneurial organizations have developed an intense activity and have had also an effective impact. However, the participation of labor unions was almost completely ignored in the beginning and is today only in its incipient phase.[19]

Thus, there existed a marked disequilibrium between the participation of the entrepreneurial and the labor sector.[20]

Among these "privileged sectors," entrepreneurs were the most active; they were organized in Chambers of Industry[21] through which their demands were channelled. (This does not mean that traditional personal contacts—considered more fruitful—were abandoned; it is only that here we are dealing with concrete instances of organized participation in the process.) The main recipients of these organized entrepreneurial demands were the governments. The process of tariff renegotiation—the most important means of industrial protection and, at the same time, the principal field of entrepreneurial interest—illustrates this situation. Tariff renegotiations took place in the Executive and Economic Councils, and it was common practice for the owner or representative of the industrial plant whose protection was at stake to participate as member of the national delegation of the country where the plant was located—as an "advisor" to governmental representatives, who negotiated and decided the levels of protection demanded by the entrepreneur as if they were the "national interest" of the country where the plant was located. A study on entrepreneurial attitudes has concluded that "the Common Market has introduced in Central America a situation in which business and governmental interests coincide much more than before, since governments are trying to grant to the business community substantial incentives and privileges."[22]

The governments supported the entrepreneurial demands at a regional level as a means of obtaining the allegiance of the entrepreneurs at an internal level. Private entrepreneurs had a decisive influence on the unstable existence of governments in the area and, consequently, on the national bureaucracies. This situation acted against the transfer of expectations to the regional entities since the entrepreneurial demands were satisfactorily rewarded by the governments. The existence of a regional federation of chambers of industry (FECAICA), formed by the five national chambers, did not alter the situation since it acted at a very general level without arriving at a regional articulation of its members' interests.[23]

In addition to this general satisfaction of the interests of the national entrepreneurial groupings by government action at the regional level, there is the question of the role of the regional institutions within this restricted and dependent framework. In fact, their role was limited to the gathering of information and the organization of the meetings where governmental officials joined with private interest representatives, in well differentiated national delegations, to make the ultimate decisions. In this context, the technocrats, who depended on governmental contributions for their existence, cautiously avoided advancing those initiatives which tended to orient the demands of private interest groupings toward the regional institutions, this for fear of the consequences such action would have on their futures. A large proportion of those who worked for regional institutions were recruited from executive positions in the national bureaucracies. In the regional institutions, they enjoyed relatively greater stability and comparatively better employment conditions—higher salaries, tax exemptions, diplomatic protection, and a higher social status. The tacit agreement between the technocrats and the bureaucrats not to threaten their own existences, tended to limit the active role that theoretically was assigned to the technocrats and thereby stultify the process.[24]

The Pattern of Outcome of the
Decision Making Process

In our general analysis of institutional autonomy it is important to observe the pattern of outcome of the decision making process. Two cases or "autonomous policy contexts,"[25] selected according to their importance will be studied here: balanced development and the solution of free trade conflicts.

Balanced Development. Observers generally agree that one of the most important problems confronting integration in developing countries is related to the distribution of the benefits among the participants.[26] Some of the most prominent theoreticians of integration have seen in "unsatisfactory bilateralism and reciprocity ... the ideological justification for the inauguration of a different procedure for aiming at general satisfaction."[27]

The idea that economic integration in Central America had to be beneficial to all the participants—as pointed out in Chapter 2—was one of the requisites that ECLA had put forth from the beginning as fundamental to the launching and development of the integration program. ECLA understood reciprocity, within the limited goal of planned industrialization, as assuring each participant that one of each of the five industries selected would be located on its territory. Probably because the differences in degrees of development among the Central American countries were not considered very pronounced at that time, reciprocity was defined only as the equitable distribution of benefits. But even the less pronounced heterogeneity existing in Central America imposed an additional element on this requisite, known in ECLA's terminology as balanced development. This additional element sought to avoid in Latin America a repetition of the international division of labor between industrialized countries on one side and developing ones on the other, with the resulting damages to the latter. To achieve this purpose, it was necessary to adopt measures tending

to the achievement by the relatively less developed, within the context of integration, of a rate of growth that is not only greater than the one they would have without it, but also greater than the one achieved by the other participant countries in the process, in order that the differences in development that separate these last from the first tend to be reduced.[28]

Consequently, it was not enough to assure all the participants that they would receive the same benefits; it was necessary that the relatively less developed obtain more than the others to reduce the differences existing among them, briefly, that the relatively less developed members should receive preferential treatment.

How was the principle of balanced growth applied in Central America?

In June 1964 an extraordinary meeting of the Economic Council took place to study the problem presented by the refusal or delay of some members in ratifying seven treaties dealing with the program of economic integration—Honduras had not ratified six, nor Nicaragua four, nor Costa Rica and El Salvador three. At the meeting, representatives of the regional federation of chambers of industry urged their immediate ratification. As a result the Economic Council requested an evaluation of the whole program from ECLA's México Office to determine the causes of the delay and the changes that would be required once the treaties were ratified.[29] In September of the same year, the Joint Planning Mission issued a set of guidelines for the adoption of a regional development policy, in which, by comparing the growth of per capita income among the participants, it concluded that it was "indispensable to grant to Honduras preferential treatment in external financing and in the location of certain productive activities of great dynamism."[30] In December, the Honduran Association of Industrialists suggested that its government not ratify and renegotiate the agreement for the unification of fiscal incentives to industrial development "to adequate it to the policy of balanced development."[31] Thus, it was obvious that one of the reasons for Honduras delays in ratifying the integration treaties was the increasing dissatisfaction with its relative position in the program. And that the immediate cause for this situation was found in the relatively few number of new industries that it had been able to attract.[32] Consequently, Honduras would not ratify the agreement to standardize fiscal incentives in the region and the other treaties until the other members would agree on granting it preferential treatment that, by acknowledging its relatively weaker position, would permit the use of differentials in fiscal incentives to attract new industries to its territory. Actually the agreement on fiscal incentives already recognized this by granting to Honduras and Nicaragua the concession that industries located on their territories would enjoy these incentives for longer terms—two years more for Honduras and one year for Nicaragua—than those located in the other member countries (Transitory Article V of the Agreement). But this was considered insufficient to guarantee the industrial growth of the relatively less developed areas. In October 1965, ECLA circulated the evaluation requested two years before. This document recognized Honduras' disadvantageous situation and in a separate document proposed a set of measures to alleviate it. One of these measures was the granting of wider margins to Honduras in the use of fiscal incentives.[33] The evaluation and the measures were discussed at the meeting of the Committee of Economic Cooperation convoked specifically for this purpose in Guatemala in January 1966, and a set of recommendations was adopted accepting ECLA's suggestions.

The recommendations to the Economic Council dealt with the selection of those industries that could be located in Honduras within the Regime of Integration Industries and with the granting of preferential treatment in the use of fiscal incentives. It was recommended that the Central American Bank, the UN, and other international lending institutions give priority to the financing of projects located in Honduras. In exchange, Honduras was asked to ratify those treaties that it had not yet approved.[34] Following these recommendations, a protocol to the agreement for the standardization of fiscal incentives was signed in September 1968 by the Economic Council, by which preferential treatment was granted to Honduras.

The results of this decision can be placed between two of the lower degrees of decision-making patterns described by Haas—the minimum common denominator and the splitting of the difference.[35] To the first belongs that part of the decision that satisfied the demands of Honduras, which had used such unfriendly means as postponing the ratification of other treaties or disguised threats of withdrawal to exercise pressure on the other participants. To the second belongs the role of ECLA's México Office in solving the problem, and to a lesser extent that of the Joint Planning Mission, in suggesting the measures that were finally adopted.

In balanced development observers of integration in Latin America saw one of the means of transcending the limited and dependent institutional situation, because by using this problem, it was hoped that the regional technocrats could exploit "the growing but limited convergence of the major powers and the special fears and hopes of the smaller ones"[36] to arrive at a degree or level of integration in which decisions were adopted by "upgrading the common interest." But the making of the decision in Central America did not work as it theoretically should. This is particularly evident if one observes that the role of the regional technocrats was limited to suggesting those measures that gave satisfaction to the dissenting member, but in no instance would they try to transform the institutional setting, since their concern was to preserve the existing structure. That is, the crisis was perceived by the technocrats as one that imperiled the existence of the program as it was; the possibility of their obtaining wider powers—for instance, to apply the development program done by the Joint Planning Mission to precisely give satisfaction to Honduras—was never envisaged. In fact, ECLA's México Office assured the participants that the measures proposed referred to the existing treaties and institutions, did not require the implementation of new actions nor demand a large proportion of the members' resources, and that they were meant only to create new opportunities for the less developed member.[37] Furthermore, the role of the Central American institutions, particularly SIECA, was limited to bringing the participants together, once they had reached general agreement on the need of signing the protocol, to approve the recommendations already adopted by the Ministers of Economic Affairs in the Committee of Economic Cooperation.[38] From this one assumes that the technocrats were not interested in directing the program nor in promoting those activities that would cast them in a more active, though less secure, role. Nevertheless, it can be contended that for the immediate future the program in Central America did not require further measures, and that all that was needed was the application of those instruments already approved, but that, for one reason or another, remained unenforced. This argument—advanced frequently by regional technocrats—confirms the assertion that regional institutions were devoted to the solution of those questions raised at the restricted and dependent level mentioned above, which in itself prevented the successful use of those opportunities to transcend it. Instead, as the analysis of the application of the principle of balanced growth has hopefully demonstrated, these opportunities were used to reinforce the *status quo*.

Conflict Resolution of Free Trade Problems. Because free trade was the most successful activity of the program for regional economic integration, its application provoked frequent conflicts whose analysis will help to describe the attitude of the participants on more routine questions, more routine when compared to the relatively unusual situation described above.

As mentioned in Chapter 4, the most important obstacles to free trade arose from the question of the origin of the products. This, in turn, was a consequence of the competitive national policies of industrial protection and of the discrimination caused by the imposition of sales or consumption taxes by one member on goods produced in another, in an effort to avoid competition and the diminution of revenues that free trade implied. These problems appeared with such frequency that the Executive Council—where they were discussed—decided to establish a permanent point in the agenda of its monthly meetings, when, in an order decided by lottery, each country made its complaints. Their frequency can be further illustrated by the fact that during seven meetings held by this multilateral institution between October 1966 and October 1967, 115 of these questions were discussed.[39] Some of the complaints will be described here with special emphasis on the part played by the regional institutions in their solution and the degree of autonomy that this represents.

These skirmishes usually began with a complaint from a private producer to the Ministry of Economic Affairs of the country where his industry was located, a complaint about the disastrous effects that the competition of similar goods produced in other member countries was having on his trade. He argued, sometimes correctly, either that the competitor had received a higher degree of protection from his government or that his own existence was imperiled. The Minister immediately reacted to the need to protect the "national interest" by demanding of his colleague in the other member country an explanation of the situation, and accompanied his demands with retaliatory measures—such as stopping the goods at the border or imposing consumption or sales taxes on them, arguing that there were doubts as to the origin of the products or that the exporter was dumping. If the situation was not solved by these bilateral contacts between governmental representatives, where private producers actively participated, the government against whom the retaliatory measures were applied raised the question in the monthly meeting of the Executive Council. In spite of the provision of the General Treaty that the Permanent Secretariat (SIECA) was charged with overseeing the fulfilling of its terms and that free trade was one of them, in practice SIECA participated in the bilaterial discussions only if it was invited by the contenders. Most of the questions raised for the first time in the Executive Council by a country were not discussed because of the absence of a "basis for judgment." Thus, when the question was raised by the complaining member, the first decision was to ask SIECA and/or ICAITI to clarify the situation after listening to the arguments of the members involved.

At the meetings of the Executive Council, the governmental delegates were accompanied by the private producers affected, who were "advisors" to the national delegations. And although they did not vote, because this was a privilege of the "Vice-Ministers" of Economic Affairs, they intervened actively in the discussion of the problem, while the officials limited themselves to defending the "national interest." The problem was then turned over to SIECA or ICAITI or both so that they could study the situation and note the violations to existing treaties. These studies frequently had to go into technical questions (for example, the origin of the products) and took considerable time to complete. In any case, at least one month had to pass before the question could be further discussed at the following meeting of the Executive Council. In that time, free trade was interrupted, and relations between the members involved were further embittered since the retaliatory measures persisted or were escalated; efforts at bilateral solutions were put aside while both parties awaited the results of the investigation by the permanent institutions.

Once the report was available, it was discussed by the governmental delegates and private producers in the Executive Council. Generally the judgment of the permanent institutions gave ground to a resolution requesting the delinquent member to abolish the discriminatory measures.[40] This resolution was adopted by a majority of four with a negative vote from the country affected, but given the absence of sanctions if a member refused to apply the resolution, the question appeared frequently at the next meeting where the reluctant government produced excuses for not abiding with the resolution. Finally, the problem was usually settled when the member government, judged to be at fault, voluntarily agreed to apply the resolution, either because it had bilaterally exchanged concessions with the other on retaliatory measures or because the pressure of the multilateral meeting made it impossible to continue with its posture without provoking a major crisis. The multilateral decision allowed the government concerned to dismiss the protectionist demands of the private producer with more authority, after having supported them unsuccessfully throughout the discussion. But if then the question was not solved, the complaining member could raise the question in the Economic Council at a ministerial level. These meetings usually supported the decision adopted at the lower level, which explains the very few instances in which this appeal was made and also the absence of the extreme possibility of a demand for arbitration.

The frequency with which these "small fires" appeared produced regulations that formalized the practices described above, including the passive, consultative role of the permanent institutions.[41] This meant that numerous cases continued to be referred to the Executive Council and continued to be solved slowly; free trade was constantly interrupted and the protagonists continued to exhibit a lack of integrative behavior.

From the observation of these more frequent conflicts, one can see that the main participants in the decision-making process were the national bureaucrats, the private producers, and the regional technocrats, the latter as consultants. And from the contenders' attitudes, it is possible to conclude that each tried to satisfy its own interests by "splitting the difference," since the questions were ultimately solved by exchanging concessions. Finally, whether or not the decision would be honored was left to the will of the delinquent government, who frequently postponed action to obtain more concessions or to justify dismissing the protectionist demands of the private producers because of the possibility of greater retaliation or more severe consequences resultant from the other members' impatience.

6 Conclusions

What are the constant elements that permitted the establishment and evolution of the Central American integration process? How well did its institutions fulfill the expectations of theoreticians and practitioners?

Central America's economic integration process, during the period under study here, was characterized by the avoidance of "high" costs for the participants. This notion of "high" and "low" costs refers to the efforts and sacrifices—material and political—that the countries had to make to implement integrative measures. It is related to the importance to the members' national economies of the sectors affected by integration as well as to the degree of controversy aroused by the implementation of the measures adopted in each sector. Finally, it assumes that the process' relevance is proportionate to the sacrifices required of the participants and, consequently, to the degree of controversy that the process generates.[1]

The avoiding of the sacrifices and costs involved in implementing integrative measures affected the emergence of the larger political entity from the integrative process and, probably of more importance, the modernization of Central America. This constant avoidance of sacrifice explains why we had to look at the influence of external factors on the origins and development of Central American integration.

ECLA's participation assured the member governments that the costs of the necessary preliminary studies would be financed by the United Nations Technical Assistance Administration. By suggesting import substitution as the solution to their backwardness, it also assured them that something was going to be added to what already existed without requiring major transformations in their societies, as would have been the case if the experiment had started with the agricultural sector of their economies. It can be argued that these limited goals were proposed as a device for obtaining a preliminary consensus that would allow, with the passing of time, a gradual application of more relevant measures in more controversial sectors. This argument can be supported by the fact that these preliminary measures were in line with the reformism cherished by the ruling elites to which they were being proposed.

But further sacrifices to achieve more daring results were avoided with equal assiduousness. Thus, the signing of the General Treaty of 1960, by which free trade became the general rule, was facilitated by the concrete offer of the United States government to alleviate the financial consequences. It should be recalled that the integrative process had arrived at a deadlock—ECLA was going to reduce its participation, and the members were being asked to increase their contributions because the United Nations Technical Assistance Administration did not have the funds. Nevertheless, just at this moment, the process moved into a more advanced stage because the participants were assured that their powerful neighbor would mitigate the costs and sacrifices that these relatively more audacious measures entailed. The fact that their adoption meant fulfilling some requisites demanded by the United States—such as abandoning some of ECLA's recommendations, e.g., the Regime of Integration Industries, or the exclusion of ECLA's México Office, as in the signature of the Tripartite Treaty—seemed to have no bearing on these decisions.

The actual operation of the program shared this characteristic. One has only to observe the dependence of regional institutions on external resources, or to look at the number of projects that remained on paper because of the lack of foreign funds to finance them.

The question which follows from the preceding considerations is: what is the degree to which this situation affected the results expected by theoreticians and practitioners?

The avoidance of "high" costs that characterized the Central American integrative experience affected the gradual emergence of a larger political entity among the participants. Instead of learning to upgrade the common interest, the members learned to put in practice those measures whose economic consequences could be lessened by foreign assistance and that, consequently, demanded a minimum of sacrifices from them. Instead of the gradual transfer of expectations to a larger entity, each participant pursued the satisfaction of his individual interests by means of very uncooperative methods, such as retaliatory measures against its partners or the zealous protection of its national producers from regional competition. In other words, the process became one in which short-term benefits were more important than any long-term expectation. The outcome was very different from the one expected by the theoreticians, and it is important to try to determine the reasons for such a situation.

The theoreticians expectations seemed very optimistic when one looks to the Central American reality—the transitional stage that Latin America was assumed to be going through, which was characterized by the existence of interest groups clamoring for attention. The actual participants in regional activities were those traditional groups that held the levers of power in the member countries and who were able to articulate their demands in an organized way. This is in large part due to the fact that the emphasis on industrialization excluded the largest sectors of their populations. Even within the industrialized sector, other less well-organized forces—such as the labor organizations—remained excluded from regional activities. The setting was then very different from the fertile environment seen by theoreticians as necessary for the successful operation of their propositions.

The same is true of the *técnicos*. Although they occupied the strategic position in the process of change, their position was weak and dependent. Thus, instead of supranational agents devoted to upgrading the common interest by overcoming the built in limitations in their tasks, they became more concerned with the preservation of their weak but prestigious existence. They knew that if they dared to organize those groups that remained excluded from the integrative process—labor organizations—so that they would participate and be used against those within the game—the industrialists—their existence would be imperiled. So they devoted themselves to preparing technical studies and putting out "small fires" and thus avoiding the advancement of propositions that could lead to a direct confrontation with those groups on whose support their existence depended. The preservation of the regional institutions, where they were entrenched, constituted the main activity of the *técnicos*. The process lost its character as an instrument for building a larger entity among the participants or for their modernization; it became an end in itself. The *técnicos* gave up their role as agents of change and became agents of the *status quo*. Regional institutions depended for their functioning on contributions from the member governments and on foreign sources. The fact that the Central American governments exhibited a persistent reluctance in fulfilling their financial commitments to the budgets of regional institutions—which made for a state of almost perpetual financial crisis—forced these institutions to look for other more dependable sources abroad. Experience shows that there are some costs of integrative measures that are not financial. These are their "political" costs, usually understood as the transfer of powers to the regional institutions. There were cases, such as the telecommunications network, in which—despite the availability of foreign financing to implement the project—the government refused its approval until the provision requiring a transfer of powers was deleted. The United States government, the main source of financing for regional activities, required the fulfillment of certain requisites before granting funds, and in line with its suspicious attitude toward ECLA's activities, refused to support the Regime of Integration Industries. The application of ECLA's reciprocity formula depended on the success of the Regime of Integration Industries. But experience had taught the Central Americans that their powerful North American neighbor was able to hinder the workings of those measures with which it did not agree and they preferred to keep integration within the limits allowed. In this last sense, the situation was similar to that of supporting the traditional oligarchical interests, since both the powerful neighbor and these groups, those elements on which the process's existence depended, could and did impede the adoption of those measures that could contribute to the politicization and relevance of the regional institutions, and that implied at least a decrease in their influence, if not actually going against their interests. It is not strange at all that, given these conditions, Haas and Schmitter concluded that only a crisis whose solution cannot be achieved through foreign assistance would permit the politicization of economic integration in the Hemisphere.[2] Schmitter went on to conclude that in Central America, instead of the spill-over process on which politicization depends, integration spilled around.[3]

Accepting the imposed limits affected not only the emergence of a larger political entity, perceived by the Central Americans as a very long-term goal, but also the more urgent question of modernization, viewed as integration's most important goal. The impact of the process in the participants' modernization was limited, if not negative, because of the emphasis on avoiding "high" costs by import substitution.

Agriculture, by far the largest sector of the members' economies, was affected only marginally and remained excluded from the integration process, principally because the solution of the problems that surrounded it required not only massive financial resources, but a direct confrontation with the most conservative and powerful vested interests existing in the five countries. Those responsible for integration in Central America seemed to have forgotten that "if industrial integration is to be a key element in a policy of industrial development, it is necessary to tackle also the problem of agricultural improvement";[4] in fact they tried to avoid the whole issue. But the expansion of the national markets—the main justification for economic integration among less developed countries—was not only related to the number of consumers that could be brought together but also to the more controversial question of the income level of the majority of the participants' populations. The creation of a regional market of 15 million people in Central America was an illusion as long as the capacity to consume of the peasant sector was not drastically improved. To this extent any policy concerned almost exclusively with industrialization was severely limited and risked becoming a costly exercise in exaggerated protectionism. The process of economic integration in Central America, by staying within the restricted limits set up by import substitution, was also likely to produce negative consequences for the participants' economic growth to the extent that it protected inefficient activities. This began to happen in the area, when the member governments made protection of local industries their most important goal.[5]

The conclusions that can be derived from these considerations are that (1) "low" costs produced insignificant and negative results; (2) foreign aid was not always useful, particularly in more relevant sectors; and (3) the successful adoption of preliminary measures requiring few sacrifices in noncontroversial sectors did not assure that relatively more relevant activities in other sectors could be undertaken with the same ease.

Economic integration in Central America became, as Haas and Schmitter had warned, the prisoner and victim of the conditions that permitted its existence.[6] Nevertheless, if in spite of everything the Central American technocrats are able to produce the changes urgently required in the Isthmus by using the limited means available to them, then they will be the philosopher-kings who have discovered "a magic new route to painless development."[7]

Notes

Introduction

1. See for instance, SIECA, *5 años de labores en la integración económica centroamericana* (Guatemala: SIECA, October 12, 1966).
2. So called for its relation to functionalism. For a comparison of both see: C. Pentland, "Functionalism and Neofunctionalism: Some Comments." Also Paul Taylor, "The Functionalist Approach to Integration." Papers submitted to the Conference on Functionalism and the Changing Political System, Carnegie Endowment for International Peace, November 20-24, 1969. Mimeographed.
3. Ernst B. Haas and Philippe C. Schmitter, "Economics and Differential Patterns of Political Integration: Projections about Unity in Latin America," in *International Political Communities: An Anthology* (New York: Anchor Books, 1966), p. 265.
4. Philippe C. Schmitter, "Three Neo-Functional Hypotheses About International Integration," *International Organization*, 23 (Winter 1969): 166.
5. Ibid., p. 162.
6. Haas and Schmitter, "Economics and Differential Patterns," pp. 258-299.
7. Ernst B. Haas, "The Uniting of Europe and the Uniting of Latin America," *Journal of Common Market Studies*, 5 (June 1967): 334.
8. For a discussion of the "functional equivalents" see: Ibid., pp. 338-343. They were also suggested in Haas and Schmitter, "Economics and Differential Patterns," pp. 284-293.
9. Albert O. Hirschman, *Journeys Toward Progress* (New York: The Twentieth Century Fund, 1963), pp. 227-97.
10. Haas and Schmitter, "Economics and Differential Patterns," p. 285.
11. Ibid.
12. Ibid., p. 286.
13. Haas, "The Uniting of Europe," p. 343.
14. Haas and Schmitter, "Economics and Differential Patterns," p. 296.
15. Stanley Hoffman, "Obstinate or Obsolete The Fate of the Nation-State and the Case of Western Europe," in *International Regionalism*, ed. J.S. Nye, pp. 177-230.

Chapter 1. Central America After The Second World War

1. Karl Deutsch et al., "Political Community in the North Atlantic Area," in *International Political Communities: An Anthology* (New York: Anchor Books, 1966), pp. 1-44; Ernst B. Haas, "The Challenge of Regionalism," in *Con-

temporary Theory in International Relations, ed. Stanley Hoffman (Englewood Cliffs: Prentice Hall, Inc., 1960), pp. 228-232; Haas and Schmitter, "Economics and Differential Patterns," pp. 266-68.

2. Haas and Schmitter, "Economics and Differential Patterns," p. 267.

3. Raymond Aron, *Paix et Guerre entre les Nations*, 4th ed. (Paris: Calman-Lévy, 1952), p. 155.

4. Haas and Schmitter, "Economics and Differential Patterns," p. 268.

5. Deutsch et al., " Political Community," p. 17.

6. Ernst B. Haas, "International Integration: the European and the Universal Process," in *International Political Communities*, p. 95.

7. UN-ECLA, *Evaluación de la integración económica en Centroamérica* (E/CN.12/327/Rev. 2), New York, 1966, p. 7.

8. UN-ECLA, "Central American Post-War Exports to the United States," *Economic Bulletin for Latin America*, 5 (October 1960): 25.

9. Joseph Pincus, "The Central American Common Market," (US Department of State, Agency for International Development, Regional Office for Central America and Panama Affairs ROCAP, September, 1962), p. 54. Mimeographed.

10. Organización de Estados Americanos, Secretaría General, "Estudio Económico de América Latina, Parte 2: Algunos aspectos de la producción y el comercio de América Central" (Washington, D.C., 1964), pp. 342-58. Mimeographed.

11. UN-ECLA, *La política tributaria y el desarrollo económico en Centroamérica* (E/CN.12/486), September 1956, pp. 79-108.

12. Jean Siotis, "The Secretariat of the United Nations Economic Commission for Europe and European Economic Integration: The First Ten Years," *International Organization*, 19 (Spring 1965): 179.

13. IBRD-IDA, "Economic Development and Prospects of Central America," vol. 3: "Agriculture" (Report Number WH-170a), June 5, 1967, p. 2. Mimeographed.

14. Ibid., p. 17.

15. This term is used here as Schmitter and Haas use it in "México y la Integración Económica Latinoamericana," *Desarrollo Económico*, 4 (July-December 1964): 125.

16. Studies on Central American comparative politics are scarce. The only study the author was able to find was: Charles W. Anderson, "Politics and Development Policy in Central America," in *Latin American Politics*, ed. Robert D. Tomasek (New York: Anchor Books, 1966), pp. 544-565. This article analyzes three countries: Costa Rica, El Salvador and Guatemala.

17. The treaty rules on cases of legitimate defense and spells out the assistance that should be furnished to the state attacked. For the text of the treaty see, Instituto Interamericano de Estudios Jurídicos Internacionales, *El Sistema Interamericano* (Madrid: Ediciones Cultura Hispánica, 1966), pp. 473-79.

18. ICJ, *Reports of Judgments, Advisory Opinions and Orders*, Judgment of November 18, 1960, p. 192.

19. UN-ECLA, "Análisis y perspectivas del comercio intercentroamericano 1934-38 a 1946-52" (E/CN.12/CCE/10), September 11, 1954. Mimeographed.

20. UN-ECLA, "El Transporte en el Istmo Centroamericano" (E/CN.12/356; ST/TAA/Ser.C/8), México, September 1953, p. 165.

21. This monopoly was abolished in 1958 by the United States government (*US v. UFCO*, settled by consent in 1958). "The US Government charged UFCO with restraining and monopolizing interstate and foreign trade in bananas in violation of the Sherman and the Wilson Act. Among the practices specifically cited were: a) the acquisition of nearly all banana land in Central America; b) *the control of port and communications facilities used in banana shipment* [italics mine]; c) pricing activities to drive out competitors; and d) the regulation of supply so as to coerce customers." US Congress, Senate. Committee on Foreign Relations. Subcommittee on American Republics Affairs, *United States and Latin American Policies Affecting their Mutual Relations*, a study prepared by the National Planning Association. January 31, 1960 (Washington: Government Printing Office, 1960), p. 112.

22. US Congress, Special Committee Investigating the National Defense Program, "Inter-American Highway," 80th Congress, First Session, Report No. 440, 1947, quoted in John F. McCamant, *Development Assistance in Central America* (New York: Frederick A. Praeger, Inc., 1968), p. 100.

23. UN-ECLA, *El Transporte*, p. 191.

24. Ibid., p. 149.

25. Ibid.

26. IBRD-IDA, *Transportation*, 6, p. v.

27. UN-ECLA, *El Transporte*, p. 1.

28. The data used here, unless noted otherwise, has been taken from: UN-ECLA, "Situación y tendencias demográficas recientes en Centroamérica" (E/CN.12/CCE/356; TAO/LAT/86), May 15, 1968. Mimeographed. UN-ECLA, "Distribución de la Población en el Istmo Centroamericano" (E/CN.12/CCE/357; TAO/LAT/87), August 1, 1968. Mimeographed.

29. UN-ECLA, "Situación y tendencias demográficas recientes en Centroamérica," pp. 34-46.

30. A description of the formalities that a citizen of El Salvador had to fulfill to leave his country can be found in: UN, *Transportes interiores en El Salvador* (ST/TAA/J. El Salvador/RII), New York, 1952, p. 92.

31. UN-ECLA, "Aspects of the Interrelations Between the Trends of Economic Development and Human Resources in México, Central America, and Panama" (Preliminary version) (CEPAL MEX/68/14), August 12, 1968. Mimeographed.

32. Philip J. Dick, "The Role of the Mass Media in Central American Development," 1968, p. 52. Studies of Entrepreneurial Behavior in Central America, Project Director: Calvin P. Blair. Mimeographed.

33. J.S. Nye has emphasized the importance of external factors in the development of Central American integration in his "Central American Regional Integration," in *International Regionalism: Readings*, ed. Joseph S. Nye, Jr. (Boston: Little Brown Company, 1968), pp. 414-20. Also: Aaron Segal, "The Integration of Developing Countries: Some Thoughts on East Africa and Central America," *Journal of Common Market Studies*, 5 (March 1967): 270.

Chapter 2. The Economic Commission for Latin America (ECLA)

1. UN-ECLA, "Vigésimo aniversario de la CEPAL," *Boletín Económico de América Latina*, 13 (November 1968): 140.
2. Ibid., p. 139.
3. Raúl Prebisch, "The Economic Development of Latin America and Some of its Principal Problems," *Economic Bulletin for Latin America*, 7 (February 1962): 3.
4. Ibid.
5. Carlos Manuel Castillo, *Growth and Integration in Central America* (New York: Frederick A. Praeger, Publishers, 1966), p. 77.
6. UN-ECLA, "Resoluciones del Comité de Cooperación Económica del Istmo Centroamericano," vol. 2: "Textos" (E/CN.12/CCE/358), July 1966, p. 1. Mimeographed.
7. UN-ECLA, "Desarrollo Económico en Centroamérica" (E/CN.12/275), June 16, 1951. In "Resoluciones del Comité de Cooperación Económica," vol. 2: "Textos," p. 1.
8. UN-ECLA, "Creación de una oficina de la CEPAL en México" (E/CN.12/284), June 16, 1951, vol. 2: "Textos," p. 3.
9. Carlos Manuel Castillo, *Growth and Integration*, p. 78.
10. Article 1, Charter of the Organization of Central American States (ODECA), in *Reuniones y Conferencias de Ministros de Relaciones Exteriores de Centroamérica, 1951-1967* (Organización de Estados Centroamericanos ODECA, Publicaciones de la Secretaría General, n.d.).
11. Ibid., Article 14.
12. Ibid, pp. 14-26.
13. Nye, "Central American Regional Integration," p. 394.
14. UN-ECLA, "Actas y Documentos de la I Reunión del Comité de Cooperación Económica del Istmo Centroamericano" (E/CN.12/AC.17/1-23), Tegucigalpa, Honduras, August 23-28, 1952. Mimeographed.
15. Ministerio de Relaciones Exteriores de El Salvador, *Denuncia de Guatemala a la Carta de San Salvador y Retiro de la ODECA*, (publicaciones del Ministerio de Relaciones Exteriores de El Salvador, San Salvador, 1953).
16. ODECA, *Reuniones y Conferencias*, pp. 27-36.

17. Ibid., p. 64.

18. The most important officials of the Secretariat went to Tegucigalpa. Prebisch, Urquidi, Martínez Cabañas, Santa Cruz, Mayobre, Lara Beautell were there backing the proposals.

19. UN-ECLA, "Informe preliminar del Secretario Ejecutivo de la Comisión Económica para América Latina sobre Integración y Reciprocidad Económica en Centroamérica" (E/CN.12/AC.17/3), August 1, 1952, pp. 5-36. Mimeographed. This document contains the main guidelines of ECLA's proposal in Central America and will be hereinafter referred to as "Integración y Reciprocidad en Centroamérica."

20. Ibid., p. 37.

21. Ibid.

22. Ibid., p. 38.

23. Ibid., p. 39.

24. Haas and Schmitter in their "Economic and Differential Patterns" have emphasized the importance of the reciprocity formula in the launching of LAFTA, but consider that its main purpose was "to create a climate of trust" (p. 289).

25. UN-ECLA, "Integración y Reciprocidad en Centroamérica," p. 39.

26. Ibid.

27. At the time of the meeting, El Salvador had entered into this sort of treaty with Honduras (1916), Guatemala (1951), and Nicaragua (1951); for a study of this practice see: Antonio Collart Valle, "Política comercial, tratados de comercio y unificación arancelaria como factores en el proceso de integración económica centroamericana," in *Integración Económica de Centroamérica*, ed. Jorge Luis Arriola (San Salvador: ODECA, Publicaciones de la Secretaría General, 1959), pp. 307-344.

28. UN-ECLA, "Integración y Reciprocidad en Centroamérica," p. 42.

29. Technology was considered fundamental by the Secretariat, and a concrete proposal for the creation of a regional institute was advanced in the meeting, see: UN-ECLA, "Nota del Secretario Ejecutivo sobre integración económica y cooperación tecnológica" (E/CN.12/AC.17/4), August 4, 1952. Mimeographed. The transportation systems were being studied by a mission of ECLA and the Technical Assistance Administration, see: UN-ECLA, *El Transporte*. The other sectors were to be submitted to analysis afterward.

30. UN-ECLA, "Integración y Reciprocidad en Centroamérica," pp. 52-60.

31. The minutes which reproduced textually the interventions of the delegates can be found in: UN-ECLA, "Actas y Documentos de la I Reunión." For the texts of the resolutions see: UN-ECLA, "Resoluciones del Comité de Cooperación Económica," vol. 2: "Textos," pp. 5-23.

32. UN-ECLA, (E/CN.12/AC.17/I S57), August 28, 1952. Mimeographed.

33. UN-ECLA, "Atribuciones del Comité de Cooperación Económica de Ministros de Economía del Istmo Centroamericano" (E/CN.12/AC.17/18),

August 27, 1952. In "Resoluciones del Comité de Cooperación Económica," vol. 2: "Textos," p. 17.

34. For a good summary of the activities of these periodical meetings, see: UN-ECLA, "Breve reseña de las actividades de la CEPAL en México desde su creación en 1951 hasta mayo de 1968" (CEPAL/MEX/68/II), April 18, 1968, pp. 3-7. Mimeographed.

35. The multilateral free trade treaty was also discussed in one of these groups. See: Ibid., p. 7.

36. UN-ECLA, "Informe del Director Principal a cargo de la Secretaría Ejecutiva sobre los trabajos realizados entre la Primera y la Segunda Reuniones del Comité" (E/CN.12/AC.17/27), September 10, 1953, p. 4. Mimeographed.

37. The role of foreign aid in the development of these institutions will be studied separately. An overall description of the financial support that the UN furnished to the Central American program can be found in: UN-ECLA, "Informe del Representante Regional de la Junta de Asistencia Técnica de las Naciones Unidas para Centroamérica" (E/CN.12/CCE/330 y Add. 1), January 12, 1966. Mimeographed.

38. Those related to road transport regulated road signals, circulation, temporary importation of vehicles, and so forth.

39. The proposal can be found in: UN-ECLA, "Actas y Documentos de la II Reunión del Comité de Cooperación Económica del Istmo Centroamericano" (E/CN.12/AC.17/25-44), San José, October 13-17, 1953, p. 9. Mimeographed. And the resolution in: UN-ECLA, "Resoluciones del Comité de Cooperación Económica," vol. 2: "Textos," p. 28.

40. UN-ECLA, "Resoluciones del Comité de Cooperación Económica," vol. 2: "Textos," p. 87.

41. For the full text of the treaty see: UN-ECLA, "Informe del Comité de Cooperación Económica del Istmo Centroamericano (25 de febrero a 10 de junio de 1958)" (E/CN.12/CCE/151), June 10, 1958, pp. 32-42.

42. Ibid., p. 10. For the project presented by the Secretariat see: UN-ECLA, "Informe del Comité de Cooperación Económica del Istmo Centroamericano (30 de enero de 1956 a 24 de febrero de 1957)" (E/CN.12/CCE/103), México, June 1957, pp. 30-34. The definitive text appears in: UN-ECLA, "Informe del Comité de Cooperación Económica del Istmo Centroamericano (25 de febrero a 10 de junio de 1958)," pp. 43-44.

43. Both treaties appear in: UN-ECLA, "Informe del Comité de Cooperación Económica del Istmo Centroamericano (11 de junio de 1958 a 2 de septiembre de 1959)" (E/CN.12/CCE/184), México, December 1959, pp. 25-60.

44. Haas and Schmitter, "Economics and Differential Patterns," p. 281.

45. Sol C., Jorge, "La integración económica de Centroamérica y los programas nacionales de desarrollo económico," in *Integración Económica de Centroamérica*, p. 70.

46. UN-ECLA, "Actas y Documentos de la I Reunión Extraordinaria del

Comité de Cooperación Economica del Istmo Centroamericano" (E/CN.12/ CCE/2/Rev. 1-29), San Salvador, May 4-9, 1955, p. 13. Mimeographed.

47. This problem appeared almost from the beginning, when Nicaragua opposed the location of a pulp and paper plant in Honduras, arguing that the mission that did the feasibility study did not evaluate fully the resources of both countries, in: Ibid., pp. 35-38.

48. UN-ECLA, "Informe del Comité de Cooperación Económica del Istmo Centroamericano (11 de junio de 1958 a 2 de septiembre de 1959)," p. 6.

49. UN-ECLA, "El programa de integración económica de Centroamérica," *Boletín Económico de América Latina*, 4 (October 1959): 7.

Chapter 3. The United States Government

1. For a comprehensive analysis of this period, see: US Congress, Senate, *United States and Latin American Policies Affecting their Economic Relations*, a study prepared by the National Planning Association at the request of the Subcommittee on American Republics Affairs of the Committee on Foreign Relations (Washington, D.C.: Government Printing Office, 1960).

2. McCamant, *Development Assistance*, pp. 25, 33.

3. UN-ECOSOC, *Official Records*, 3rd year, 6th session (Lake Success, N.Y., February 2 to March 11, 1948) p. 90.

4. US Congress, Senate, *United States and Latin American Policies*, p. 131.

5. James D. Cochrane, "US Attitudes toward Central American Integration," *Inter-American Economic Affairs*, 18 (Autumn 1964): 75.

6. This condition and the five following correspond to the requirements of the United States, as quoted in: US Congress, Senate, *United States and Latin American Policies*, p. 64.

7. Ibid.

8. Ibid.

9. *The Central American Common Market*, pp. 115-16.

10. *United States and Latin American Policies*, p. 64.

11. Ibid.

12. Ibid.

13. Some information on the visit can be found in: US Congress, House, *Central America: Some Observations on its Common Market, Binational Centers, and Housing Programs*, Report of Hon. Roy H. McVicker of the Subcommittee of Inter-American Affairs, Committee on Foreign Affairs (Washington, D.C.: Government Printing Office, 1966), pp. 23-24.

14. Ibid., p. 24.

15. Council on Foreign Relations, *Documents on American Foreign Relations, 1959*, ed. Paul E. Zinner (New York: Harper & Row for the Council on Foreign Relations, 1960), pp. 508-512.

16. US Congress, House, *Central America: Some Observations on its Common Market*, p. 24.

17. The possibility of immediate establishment of permanent regional institutions must have acted as an incentive for the local technocrats. For the full text of the treaty, see: SIECA, *Convenios Centroamericanos de Integración Económica*, vol. 1 (Guatemala 1963), pp. 49-53.

18. US Congress, House, *Central America: Some Observations on its Common Market*, p. 24.

19. Nye, "Central American Regional Integration," p. 417.

20. Carlos Manuel Castillo, *Growth and Integration in Central America*, p. 89. The author was Executive Director of ECLA's México Office and Secretary General of the program of Central American economic integration.

21. UN-ECLA, "Informe de la Segunda Reunión Extraordinaria del Comité de Cooperación Económica del Istmo Centroamericano" (E/CN.12/CCE/210), May 9, 1960. Mimeographed.

22. UN-ECLA, "El programa de integración económica de Centroamérica y el Tratado de Asociación Económica suscrito por El Salvador, Guatemala y Honduras: Interrelaciones y posibles formas de consolidar y acelerar la integración económica del Istmo" (E/CN.12/CCE/212), May 6, 1960, p. viii. Mimeographed. This document contains ECLA's position regarding the signature of the Tripartite Treaty.

23. Ibid., p. x.

24. UN-ECLA, "Informe del Comité de Cooperación Económica del Istmo Centroamericano (3 de septiembre de 1959 a 13 de diciembre de 1960)" (E/CN.12/CCE/224), México, February 1961, p. 6. A description of Costa Rica's attitude on its participation can be found in: Banco Central de Costa Rica, "Participación de Costa Rica en la Integración Centroamericana: Consideraciones Generales" (Departmento de Estudios Economicos EE/331/RNM:ASC/mn:ad), August 1961. Mimeographed. Also: S. Staley, "Costa Rica and the Central American Common Market," *Economia Internazionale*, 15 (February 1962): 117-30.

25. UN-ECLA, "Orientaciones básicas para el proyecto de Convenio Centroamericano de Integración Económica acelerada" (E/CN.12/CCE/211), April 28, 1960. In: "Resoluciones del Comité de Cooperación Económica," vol. 2: "Textos," p. 213.

26. UN-ECLA, "Informe del Comité de Cooperación Económica del Istmo Centroamericano (3 de septiembre de 1959 a 13 de diciembre de 1960)," pp. 53-56.

27. US-Department of State, Press Release 627, November 3, 1960. *Department of State Bulletin*, 43 (November 21, 1960): 782-83.

28. Arthur M. Schlesinger, Jr., *A Thousand Days* (London: Mayflower Books, Ltd., 1965), pp. 193, 228-29.

29. Cochrane, "US Attitudes," quoting a letter from an official of the US Department of Commerce, p. 83.

30. Ibid., pp. 83-85, quoting US-Department of State-AID "Comments on the Regime of Integration Industries of the Central American Common Market" (Washington, D.C., August 1963).

31. Miguel S. Wionczek, "Integración Económica y Distribución Regional de las Actividades Industriales (estudio comparativo de las experiencias de Centroamérica y el Africa Oriental)," *El Trimestre Económico*, 33 (July-September 1966): 470-87.

32. Raymond F. Mikesell, "El Financiamiento Externo e Integración Latinoamericana," in: *Integración de América Latina: experiencias y perspectivas*, ed. Miguel S. Wionczek (México: Fondo de Cultura Económica, 1964) pp. 214-15.

33. UN-ECLA, "Financiamiento para desarrollo e integración económicos del Istmo Centroamericano" (E/CN.12/AC.17/12), August 27, 1952. UN-ECLA, "Financiamiento del desarrollo económico" (E/CN.12/AC.17/37), October 16, 1953. UN-ECLA, "Aplicación del Régimen Centroamericano de Integración Industrial" (E/CN.12/CCE/189), September 1, 1959. All these resolutions appear in: "Resoluciones del Comité de Cooperación Económica," vol. 2: "Textos," pp. 10, 32, 188. The first study on this question was done by: UN-ECLA, "Estudio preliminar sobre problemas de financiamiento del desarrollo económico y la integración en Centroamérica" (E/CN.12/AC.17/30), September 10, 1953. Mimeographed.

34. *Department of State Bulletin*, 45 (July 10, 1961): 83-84.

35. US Congress, House, *Central America: Some Observations on its Common Market*, p. 24.

36. Banco Centroamericano de Integración Económica (BCIE), *Primera Memoria de Labores: Año 1961-1962* (Tegucigalpa, August 13, 1962), p. 41.

37. US Congress, House, *Central America: Some Observations on its Common Market*, p. 24.

38. Cyrus Frank Gibson, "Regional Foreign Aid in Central America: The Implications," December 18, 1964, Studies of Entrepreneurial Behavior in Central America, Project Director: Calvin P. Blair, pp. 11-17. Mimeographed.

39. For instance, ROCAP pays the salaries of some officials of regional institutions, with the condition that the candidate is not "communist." See Nye, "Central American Regional Integration," pp. 418-19.

40. Cochrane, "US Attitudes," quoting a letter from an AID official, p. 89.

41. Costa Rica's attitude was rather ambiguous because in August 1961 it entered into a free trade treaty for a limited list of products with Panama and Nicaragua. Costa Rica's ratification of this treaty was deposited only one month after it had adhered to the General Treaty. The same is also true of Nicaragua, who despite the fact that it had signed the General Treaty in December 1960, entered into the treaty with Costa Rica and Panama and ratified it on the same date that Costa Rica did. For the text of this treaty see: "Tratado de Intercambio Preferencial y de Libre Comercio entre las Repúblicas de Panamá, Nicaragua y Costa Rica," *Revista de Economía*, año 5, no. 1 (August 1963), pp. 1-5.

42. UN-ECLA, 'Informe del Comité de Cooperación Económica del Istmo Centroamericano (14 de diciembre de 1960 a 29 de enero de 1963)" (E/CN.12/CCE/303/Rev. 1), August 1963, pp. 47-73.

43. US Department of State, Press Release 536, September 6, 1962. *Department of State Bulletin*, 47 (September 24, 1962): 450-51.

44. US Department of State, Press Release 145, March 20, 1963. *Department of State Bulletin*, 48 (April 8, 1963): 515-17.

45. Ibid., p. 517.

46. US Department of State, "Transcript of an Interview with Secretary Rusk and Representatives of the General Federation of Women's Clubs," *Department of State Bulletin*, 48 (May 6, 1963): 639.

47. Article XVII of the General Treaty. Its full text can be found in: UN-ECLA, "Informe del Comité de Cooperación Económica del Istmo Centroamericano (3 de septiembre de 1959 a 13 de diciembre de 1960)," pp. 13-50. An author that has direct access to US official sources has declared that the inclusion of this article in the General Treaty is labeled by US officials "the crime of Managua." James D. Cochrane, *The Politics of Regional Integration: The Central American Case* (The Hague: Martibus Nijhoff, 1969), p. 209.

48. "Inter-American Economic and Social Council Reviews the Alliance for Progress," *Department of State Bulletin* (December 28, 1964), p. 900, quoted in Lloyd Dogget, "United States Policy Toward Central American Economic Integration," January 22, 1968. Studies of Entrepreneurial Behavior in Central America, Project Director: Calvin P. Blair, p. 10. Mimeographed.

49. *Department of State Bulletin*, 53 (August 23, 1965): 330-33.

50. Miguel S. Wionczek, "Latin American Integration and United States Economic Policies," in *International Organization in the Western Hemisphere*, ed. Robert W. Gregg (New York: Syracuse University Press, 1968), p. 95.

51. US Department of State, "The Presidents' Meeting at San José," *Department of State Bulletin*, 48 (April 8, 1963): 511.

52. *Gulliver's Troubles or the Setting of American Foreign Policy* (New York: McGraw Hill Book Company for the Council on Foreign Relations, 1968), p. 187.

Chapter 4. The Program of Economic Integration by Sectors of Activity

1. Leon N. Lindberg, "The European Community as a Political System: Notes Toward the Construction of a Model," *Journal of Common Market Studies*, 5 (June 1967): 344-87; and Joseph Nye, "Comparative Regional Integration: Concept and Measurement," *International Organization*, 22 (Autumn 1968): 855-80.

2. Siotis, "The Secretariat of the United Nations Economic Commission," pp. 178-79.

3. Instituto Interamericano de Estudios Jurídicos Internacionales, *Derecho Comunitario Centroamericano* (San José: Trejos Hermanos, 1968) pp. 29-32; Carlos Manuel Castillo, "Estado del Programa de Integración Económica Centroamericana," *Universidad de San Carlos, No. 71* (Guatemala, July-December 1967): 17-20; Consejo Económico Centroamericano, "Plan de acción inmediata del Programa de Integración Económica Centroamericana," in SIECA, "Acta número veinticinco, 17 Reunión Extraordinaria," Tegucigalpa, March 20-22, 1969. Mimeographed.

4. Data on free trade has been taken from: IBRD-IDA, "Main Report," 1; "Statistical Appendix to Main Report," 2; "Industry," 4; and SIECA, *Cartas Informativas*.

5. UN-ECLA, *Evaluación de la Integración Económica en Centroamérica*, p. 37.

6. IBRD-IDA, "Main Report," 1, Appendix A: "The Common Market," p. 9.

7. SIECA, "Problemas del Mercado Común" (SIECA/CEC-VII-O/D.T.3) Guatemala, July 27, 1967, pp. 3-16. Mimeographed.

8. ICAP, "Informe de la investigación sobre el funcionamiento de las aduanas fronterizas centroamericanas" (ICAP/INF/031/150/67) San José, March 1967, pp. 1-21. Mimeographed.

9. UN-ECLA, *Evaluación de la Integración Económica en Centroamérica* pp. 38-44.

10. Ibid., p. 38.

11. Ibid., p. 34; also SIECA, "La inflexibilidad del Arancel Uniforme y la Integración Económica" (SIECA/CEC-VIII-O/D.T.4) Guatemala, October 14, 1967. Mimeographed.

12. UN-ECLA, "Estado General y Perspectivas del Programa de Integración Económica del Istmo Centroamericano," *Boletín Económico de América Latina*, 8 (March 1963): 15.

13. IBRD-IDA, "Main Report," 1, Appendix A: "The Common Market," p. 11.

14. The most complete analysis that has been done on the Regime and its failure is by Miguel S. Wionzek, "Integración Económica y Distribución Regional de las Actividades Industriales."

15. SIECA, "Memorandum de la Sección de Desarrollo Industrial Integrado sobre Aspectos de Protección Arancelaria" (SIECA/CE-XXV/DT-25), San José, January 24, 1967. Mimeographed. And also SIECA, "Nota de la Secretaría sobre Coordinación del Desarrollo Industrial en Centroamérica" (SIECA/CEC-XIII.E/D.T. 10), Guatemala, September 14, 1966. Mimeographed.

16. The System was approved in a protocol to the Regime of Integration Industries where the first protectionist tariffs were established, articles 28 to 37 of the Protocolo al Convenio sobre el Régimen de Industrias Centroamericanas de Integración, in ESAPAC, *Los Instrumentos del Mercado Común Centroamericano* (ESAPAC/EXT/018/3000/65) San José, November 1965, pp. 128-31.

17. Articles 22 and 24 of the Agreement, in SIECA, *Convenios Centroamericanos de Integración Económica*, vol. 2, Guatemala, 1963, p. 15.
18. IBRD-IDA, *Industry*, 4, p. i.
19. Ibid., p. iii.
20. UN-ECLA, *El Transporte*.
21. IBRD-IDA, "Main Report," 1, Appendix A: "The Common Market," p. 6.
22. UN-ECLA, "Carreteras, Puertos y Ferrocarriles de Centroamérica" (E/CN.12/CCE/324. TAO/LAT/48) August 10, 1965, p. 19. Mimeographed.
23. UN-ECLA, *Evaluación de la Integración Económica en Centroamérica*, p. 87.
24. Comité de los Nueve, Alianza para El Progreso, "Informe sobre los Planes Nacionales de Desarrollo y el Proceso de Integración de Centroamérica" (Union Panamericana, Secretaría de la Organización de los Estados Americanos, Washington, D.C., August 1966), p. 135. Mimeographed.
25. Data on the Fund's resources was taken from: Consejo Económico Centroamericano, "Algunos aspectos del programa de integración económica Centroamericana, Exposición al Gobernador Nelson A. Rockefeller," in SIECA, "Acta Número Veintisiete, Consejo Económico Centroamericano, 19 Reunión Extraordinaria," Tegucigalpa, May 15, 1969, pp. 21-22. Mimeographed; data on loans is from Banco Centroamericano de Integración Económica, *Séptima Memoria de Labores, Ejercicio 1967/1968* (Tegucigalpa, August 9, 1968), p. 36; and the conditions of the loans were taken from IBRD-IDA, *Transportation*, 6, p. 11.
26. UN-ECLA, *Evaluación de la Integración Económica en Centroamérica*, p. 86.
27. IBRD-IDA, *Transportation*, 6, p. 6.
28. TSC Consortium, "Central American Transportation Study, Summary Report" (November 1965) pp. 105-129. (Mimeographed); Banco Centroamericano de Integración Económica, *Séptima Memoria de Labores*, p. 36; Comité de los Nueve, Alianza para el Progreso, "Informe sobre los Planes Nacionales de Desarrollo y el Proceso de Integración de Centroamérica," p. 136.
29. For data on road construction see: UN-ECLA, "Carreteras, Puertos y Ferrocarriles de Centroamérica," p. 11; on financing see: TSC Consortium, *Central American Transportation Study, Summary Report*, pp. 120-23, and also: John F. McCamant, *Development Assistance in Central America*, p. 95; and the projections appear in Comité de los Nueve, Alianza para el Progreso, "Informe sobre los Planes Nacionales de Desarrollo y el Proceso de Integración de Centroamérica," pp. 132-36.
30. ICAO, *Final Minutes and Constitutive Charter, Diplomatic Conference for the Creation of the Central American Corporation for Air Navigation Services* (Tegucigalpa, March 1960).
31. COCESNA, "Corporación Centroamericana de Servicios de Navegación Aérea," n.d. Mimeographed.

32. Forty percent of its 1966 budget was financed by US-AID, Nye, "Central American Regional Integration," p. 401.
33. IBRD-IDA, *Transportation*, 6, p. 48.
34. Ibid., p. iv; Banco Centroamericano de Integración Económica, *Hacia la Integración Física de Centroamérica*, March 1969, pp. 34-60; and UN-ECLA, "Carreteras, Puertos y Ferrocarriles de Centroamérica," pp. 58-171.
35. UN-ECLA, *Evaluación de la Integración Económica en Centroamérica*, p. 100.
36. Ibid., pp. 94-95; UN-ECLA "Red Centroamericana de Telecomunicaciones," (E/CN.12/CCE/344), January 31, 1966, in "Resoluciones del Comité de Cooperación Económica," Vol. 2: "Textos," p. 284.
37. Banco Centroamericano de Integración Económica, *Hacia la Integración Física de Centroamérica*, pp. 90-93; also Banco Centroamericano de Integración Económica, "Information on the Central American Regional Telecommunications Network Project and Its Financement," n.d. Mimeographed; Banco Centroamericano de Integración Económica, *Carta Informativa*, Año 3, No. 25 (May 31, 1969): 3-4.
38. IBRD-IDA, "Main Report," 1, p. x.
39. UN-ECLA, *Evaluación de la Integración Económica en Centroamérica*, p. 59.
40. UN-ECLA, "Informe de la Primera Reunión del Subcomité Centroamericano de Desarrollo Económico Agropecuario" (E/CN.12/CCE/318) December 9, 1964. Mimeographed.
41. SIECA, "La comercialización de los principales productos agropecuarios en el proceso de integración económica centroamericana" (SIECA/AGROPECUARIA/Informativo), Guatemala, March 16, 1967. Mimeographed. UN-ECLA-FAO, "The Central American Common Market for Agricultural Commodities," *Economic Bulletin for Latin America*, 10 (March, 1965): 23-47; Banco Centroamericano de Integración Económica, *Séptima Memoria de Labores*, p. 39.
42. Inter-American Economic and Social Council (IA-ECOSOC) and Inter-American Committee of the Alliance for Progress, "El avance de la integración centroamericana y las necesidades de financiamiento externo" (OEA/SER.H/XIV CIAP 274), July 12, 1968. Mimeographed.
43. Comité de los Nueve, Alianza para el Progreso, "Informe sobre los Planes Nacionales de Desarrollo y el Proceso de Integración Económica de Centroamérica," p. 121.
44. UN-ECLA, *Evaluación de la Integración Económica en Centroamérica*, p. 187.
45. Consejo Económico Centroamericano, "Algunos aspectos del Programa de Integración Económica Centroamericana, Exposición al Governador Nelson A. Rockefeller," pp. 55-56.
46. José Molina Calderón, "El Peso Centroamericano: Situación y Perspectivas," *Universidad de San Carlos*, No. 71 (Guatemala, July-December 1967): 31.

47. Consejo Monetario Centroamericano, *Tres Años de Compensación Multilateral Centroamericana* (Tegucigalpa, 1965).

48. Gustavo R. Duarte Fuentes, "Movimientos de Capital en Centroamérica: Alguna experiencia en la Cámara de Compensación Centroamericana y sus perspectivas," in *Compendio de los Estudios Técnicos presentados al II Congreso Centroamericano de Economistas, Contadores Públicos y Auditores*, ed. Instituto Salvadoreño de Fomento Industrial (INSAFI, San Salvador, March 1965), p. 406.

49. Jorge González del Valle, "Sistema de Pagos y Comercio Intercentroamericano," in *Integración de América Latina: Experiencias y Perspectivas*, ed. Miguel S. Wionczek, p. 316. The modifications were ratified by all the members in January 1964.

50. Consejo Monetario Centroamericano, *Hacia la Unión Monetaria Centroamericana* (San José, August 1968).

51. Consejo Monetario Centroamericano, *Tres Años de Compensación* Multilateral Centroamericana, pp. 63-68.

52. Consejo Monetario Centroamericano, *Hacia la Unión Monetaria Centroamericana*, pp. 48-56.

53. Jorge González del Valle, "Monetary Integration in Central America: Achievements and Expectations," *Journal of Common Market Studies*, 5 (September 1966): 14.

54. Consejo Monetario Centroamericano, "Exposición sobre la Balanza de Pagos Regional del Area presentada a la Primera Reunión de Ministros de Economía y de Hacienda de Centroamérica" (Antigua, Guatemala, April 4-10, 1965), p. 1. Mimeographed.

55. SIECA, "Exposición del Banco Central de Costa Rica sobre las medidas sugeridas para corregir los problemas de balanza de pagos y fiscal" (Tegucigalpa, January 6-7, 1967), p. 1. Mimeographed.

56. SIECA and Secretaría Ejecutiva del Consejo Monetario Centroamericano, "El problema de balanza de pagos y la integración económica centroamericana" (SIECA/CEC/CMCA/MH-1/D.T. 2), Guatemala, October 5, 1967. Mimeographed. And also SIECA, "Acta de la Primera Reunión Conjunta del Consejo Económico, el Consejo Monetario y los Ministros de Hacienda de Centroamérica" (Managua, November 13-16, 1967). Mimeographed.

57. SIECA, "Acta de la Segunda Reunión del Consejo Económico Centroamericano, El Consejo Monetario y los Ministros de Hacienda de Centroamérica" (San José, May 29-June 1, 1968.) Mimeographed.

58. Consejo Económico Centroamericano, "Algunos aspectos del programa de integración económica centroamericana, exposición al Gobernador Nelson A. Rockefeller," p. 54; see also Consejo Monetario Centroamericano, Secretaría Ejecutiva, "El Fondo de Estabilización Monetaria" (San José, March 26, 1969.) Mimeographed. Robert Triffin and Richard N. Cooper "Propuesta para crear un Fondo Centroamericano de Estabilización," (San José, June 1968). Mimeographed.

59. Banco Centroamericano de Integración Económica, *Séptima Memoria de Labores*, pp. 12-41; Consejo Económico Centroamericano, "Algunos aspectos del programa de integración económica, exposición al Gobernador Nelson A. Rockefeller," pp. 19-22; Banco Centroamericano de Integración Económica, *Origen Políticas y Gestión* (Tegucigalpa, n.d.), pp. 19-38.

60. UN-ECLA, *Informe del Representante de la Junta de Asistencia Técnica de las Naciones Unidas para Centroamérica*, Anexo 9, pp. 50-54.

61. NU-ILPES, "Centro América: Análisis y proyección del estrangulamiento externo en su proceso de desarrollo" (Santiago de Chile, August 1966), pp. V8-V10. Mimeographed. Banco Centroamericano de Integración Económica, "Bases para la formulación de una política regional en materia de fomento de inversiones" (Tegucigalpa, March 1965), pp. 8-12. Mimeographed. IBRD-IDA, Industry, 4, pp. 50-54.

62. An attempt to establish a regional policy for foreign investment was done in 1965 by the Ministers of Economic Affairs, but its results, besides being timid, mainly comprise a joint declaration of intent. See: SIECA, "Acta Número Quince del Consejo Económico Centroamericano" (San Lucas Sacatepéquez, Guatemala, June 19-21, 1965. Mimeographed. Also SIECA, "Nota de la Secretaría (sobre inversiones extranjeras)," (Guatemala, June 11, 1965). Mimeographed.

63. Data on bilateral foreign aid is from: IA-ECOSOC and Inter-American Committee of the Alliance for Progress, *El avance de la integración centroamericana y las necesidades de financiamiento externo*, pp. 102-112; regional official foreign assistance is made up of loans received by the Central American Bank as quoted in: Banco Centroamericano de Integración Económica, *Séptima Memoria de Labores*, pp. 13-25; data on bilateral and regional UN Assistance is from UN-ECLA, "Informe del Representante de la Junta de Asistencia Técnica de las Naciones Unidas para Centroamérica," anexo 9, pp. ii-viii.

64. IBRD-IDA, "Main Report," 1, p. 68.

65. SIECA, "Resoluciones del Consejo Económico" (Guatemala, June 30, 1967), pp. 3-5. Mimeographed.

66. Misión Conjunta de Programación para Centroamérica, "Presupuesto 1965"; "Presupuesto para 1966" (Guatemala, n.d.). Mimeographed.

67. Comité de los Nueve, Alianza para el Progreso, "Informe sobre los planes nacionales de desarrollo y el proceso de integración económica de Centroamérica," p. 27.

68. UN-ECLA, "Informe del Director del ICAITI" (E/CN.12/CCE/328), January 12, 1966. Mimeographed.

69. UN-ECLA, "Informe del Director de la ESAPAC" (E/CN.12/CCE/329), January 10, 1966. Mimeographed.

Chapter 5. The Institutional Setting

1. Dusan Didjanski, *Dimensiones Institucionales de la Integración Latinoamericana* (Buenos Aires: INTAL, 1967), pp. 87-133. Francisco Villagrán Kramer, *Integración Económica Centroamericana* (Guatemala: Editorial Universitaria, 1967), pp. 183-218.

2. Haas and Schmitter, "Economics and Differential Patterns," p. 265.

3. International secretariats and executive functions are used here as Siotis uses these terms in his "The Secretariat of the United Nations Economic Commission," p. 178.

4. SIECA, "Evaluación de la Secretaría Permanente del Tratado General de Integración Económica Centroamericana" (SIECA/CEC-VI-0/120) Guatemala, April 15, 1966. Mimeographed, p. 25. It should be noted that the Central American technocrats have always been jealous of ECLA's activities in the area, but this hardly transcends the inner life of both institutions.

5. ECLA's México Office has estimated that only 60 percent of its resources will be devoted to Central America in 1970-1971. See UN-ECLA, "Subsede de la CEPAL en México. Programa de Trabajo para 1970-1971" (CEPAL/MEX/68/15/ Rev. 1), November 13, 1968, pp. 5, 23. Mimeographed.

6. This question has been emphasized by Sidjanski, *Dimensiones Institucionales de la Integración Latinoamericana*, p. 130.

7. SIECA, "Relación entre la Corporación Centroamericana de Servicios de Navegación Aérea" (COCESNA) y la SIECA (SIECA/CEC-VII/D.T.7), Guatemala, December 10, 1963. Mimeographed.

8. Misión Conjunta de Programación para Centroamérica, "Informe sobre los resultados de la coordinación regional de los programas de inversiones públicas." (Guatemala, June 1965), p. 6. Mimeographed.

9. Misión Conjunta de Programación para Centroamérica, "Reunión interinstitucional para discutir el documento 'Bases para un programa centroamericano de desarrollo industrial' " (Guatemala, March 1965). Mimeographed.

10. Consejo Económico Centroamericano, Resolución 44 (CEC), "Organización institucional de la política comercial centroamericana," in SIECA, "Acta número vientiuno, Consejo Económico Centroamericano" (San José, August 16, 1967), p. 34-36. Mimeographed.

11. ODECA, *Reuniones y conferencias de Ministros de Relaciones Exteriores de Centroamérica 1951-1967*, pp. 212, 251-53.

12. Nye, "Central American Regional Integration," p. 409.

13. See Mario Castrillo Zeledón, "La Integración Económica Centroamericana ante los Sindicatos de Trabajadores," *La Universidad* no. 3-4 (May-August 1968): 233-41.

14. Hoffman, "Obstinate or Obsolete? ," p. 202.

15. Data on these four institutions were taken from: SIECA, "Proyecto de Presupuesto por Programas de la Secretaría Permanente del Tratado General de

Integración Económica Centroamericana (SIECA)" (SIECA/CEC-VIII-O/D.T. 2) Guatemala, October 5, 1967. Mimeographed; and its annex SIECA, "Cooperación financiera de ROCAP a Programas de SIECA," mimeographed; Banco Centroamericano de Integración Económica, "Establecimiento de un sistema de obtención de recursos fiscales para la integración centroamericana" (Informe rendido por Alvaro Magaña, Resolución no. AG-30/66), Tegucigalpa, May 1967. Mimeographed; UN-ECLA, "Informe del Director del ICAITI" (E/CN.12/CCE/ 328); UN-ECLA, "Informe del Director de la ESAPAC" (E/CN.12/CCE/329); UN-ECLA, "Informe del Representante de la Junta de Asistencia Técnica para Centroamerica" (E/CN.12/CCE/330 y Add. 1).

16. The Central American Bank sponsored a study to determine the possibilities of establishing a permanent system for obtaining financial resources for the regional institutions, as required by US-AID to grant the first loan to the Central American Fund of Economic Integration, see: "Establecimiento de un sistema de obtención de recursos fiscales para la integración centroamericana, Anexo 5, Resoluciones AG-29/66, AG-30/66 de la Asamblea de Gobernadores." Also, SIECA: "El financiamiento de las instituciones regionales de la integración económica centroamericana (Propuesta de la SIECA)" (SIECA/CEC/CMCA/MH-1-D.T.3), Guatemala, October 31, 1967. Mimeographed.

17. Data on public expenditure and employment of the central governments are from: ICAP, "Recursos Humanos, El Sector Público y su situación actual en Centroamérica" (ICAP/INV/021/500/68), San José, January 1968. Mimeographed; data on the staff of regional institutions are from: Nye, "Central American Regional Integration," p. 401; and data on governmental contributions are from: Banco Centroamericano de Integración Económica, "Establecimiento de un sistema de recursos fiscales para la integración centroamericana," p. 1-2.

18. Haas and Schmitter, "Economics and Differential Patterns," pp. 265, 291-92.

19. ILO, *La integración económica de América Latina: Problemas de participación y de política laboral* (Geneva: ILO, 1968), p. 27. The most complete analysis done on the problems of participation in the area is by Francisco Villagrán Kramer, *Integración Económica Centroamericana*, pp. 231-348.

20. The existence of the Labor Council within ODECA, besides being a subsidiary organ of an institution excluded from the "economic" program, reflects the same situation. ILO, *La integración económica de América Latina*, pp. 23-24.

21. The chambers of commerce of the five countries have also participated but less actively than those of industry.

22. Meldon E. Levine, "El sector privado y el mercado común. Reacciones de la iniciativa privada de Honduras, Nicaragua y El Salvador con respecto al Mercado Común Centroamericano" (Woodrow Wilson School of Public and International Affairs; Instituto Nacional de Administración para el Desarrollo *[INAD]*, Guatemala, August 1968), p. 60. Mimeographed.

23. FECAICA has issued statements demanding that there be more participation from the private sector in the program; that a customs union be formed, that the program of economic integration be based on the principle of free enterprise, and so forth. See, for instance, SIECA, "Resoluciones de FECAICA acerca del funcionamiento del programa de integración económica centroamericana" (SIECA/CE-XXV/D.T.26), San José, January 26, 1967. Mimeographed.

24. This can be observed, for instance, in their inability to obtain agreement from the member governments that tariff renegotiations did not require the signature of further treaties or protocols; and also in the disposition of the agreement for the unification of fiscal incentives to industrial development that leaves its administration to the national authorities for seven years and only then transfers this function to the regional authorities. In relation to this situation, it has been found that "the new and more plentiful generation of technocrats now staffing many organizations is less easily able to translate technical proficiency into power." Nye, Central American Regional Integration," p. 394.

25. Leon Lindberg, "Decision Making and Integration in the European Community," in *International Political Communities*, p. 212.

26. UNCTAD, *Trade expansion and economic integration among developing countries* (TD/B/85), August 2, 1966, p. 43; Miguel S. Wionczek, "Condiciones de una integración viable," in *Integración de América Latina: Experiencias y Perspectivas*, p. xxii.

27. Haas and Schmitter, "Economics and Differential Patterns," p. 297.

28. UN-ECLA, "Los países de menor desarrollo relativo y la integración latinoamericana" (E/CN.12/CCE/774), Guatemala, April 8, 1967, p. 7, 12. Mimeographed.

29. SIECA, "Acta número diez, Consejo Económico Centroamericano" (Tegucigalpa, January 26-28, 1964). Mimeographed.

30. Misión Conjunta de Programación para Centroamérica, "Centroamérica: Lineamientos para una política de desarrollo regional" (Guatemala, September 1964), p. 3. Mimeographed.

31. Quoted by Gert Rosenthal, "Algunas consideraciones acerca del 'desarrollo equilibrado' en el desenvolvimiento de la integración económica centroamericana" (Guatemala, Consejo Nacional de Planificación Económica, Memorandum 3-66, January 17, 1966), p. 3. Mimeographed.

32. The other country that demanded preferential treatment later on was Nicaragua, based on the chronic deficit in its regional trade balance, which attributed mainly to its less developed industrial sector. But no measures had been taken to answer this complaint. See Ministerio de Economía, Gobierno de Nicaragua, "Efectos del mercado común sobre la economía nicaragüense," in SIECA, "Acta número diecinueve del Consejo Económico Centroamericano Anexo 2." Managua, September 19-23, 1966. Mimeographed.

33. UN-ECLA, "El crecimiento económico de Honduras y el desarrollo equilibrado en la integración centroamericana," in *Evaluación de la integración econ-*

ómica en Centroamérica, pp. 198-206. Immediately afterwards Honduras demanded a revision of the agreement for the unification of fiscal incentives. See SIECA, "Acta número dieciocho del Consejo Ejecutivo del Tratado General de Integración Económica Centroamericana (San José, October 20-25, 1965), p. 14. Mimeographed.

34. UN-ECLA, "El desarrollo equilibrado de Honduras dentro de la integración económica" (E/CN.12/CCE/345), January 19, 1966, in "Resoluciones del Comité de Cooperación Económica," vol. 2: "Textos," p. 285.

35. Ernst B. Haas, "International Integration," pp. 95-96.

36. Ernst Haas affirms that this has been impossible in the case of LAFTA due to the absence of leadership and strategy on the part of the technocrats (in his "The Uniting of Europe and the Uniting of Latin America," pp. 342-43).

37. UN-ECLA, "El crecimiento económico de Honduras y el desarrollo equilibrado en la integración centroamericana," p. 205.

38. SIECA actually went as far as to submit a project of treaty to serve as the basis for negotiation, together with Honduras' project, at a special working group convoked for this purpose. But afterward, at a ministerial level, SIECA's role was limited to the writing up of alternatives based on the positions of Honduras and the other countries. See, SIECA, "Acta número diecisiete, Consejo Económico Centroamericano," Guatemala, February 1-2, 1966, Anexos 2 y 3. Mimeographed; "Acta número dieciocho. Consejo Económico Centroamericano," Tegucigalpa, April 19-23, 1966, pp. 5-13, Anexo 1. Mimeographed; "Acta número diecinueve." Consejo Económico Centroamericano, pp. 6-10, Anexo 1. Mimeographed. This last covers the meeting that approved the protocol.

39. SIECA, "Acta no. 23. Consejo Ejecutivo del Tratado General de Integración Económica Centroamericana," San Salvador, October 13-20, 1966, pp. 7-25. Mimeographed; "Acta no. 24. Consejo Ejecutivo del Tratado General de Integración Económica Centroamericana," Guatemala, November 15-25, 1966, pp. 7-43. Mimeographed; "Acta no. 25. Consejo Ejecutivo del Tratado General de Integración Económica Centroamericana," San José, January 25-February 3, 1967, pp. 10-37. Mimeographed; "Acta no. 26 Consejo Ejecutivo del Tratado General de Integración Económica Centroamericana," San Pedro Sula, March 9-16, 1967, pp. 13-54. Mimeographed; "Acta no. 27 Censejo Ejecutivo del Tratado General de Integración Económica Centroamericana," San José, July 20-30, 1967, pp. 13-54. Mimeographed; "Acta no. 29 Consejo Ejecutivo del Tratado General de Integración Económica Centroamericana," Tegucigalpa, September 18-23, 1967, pp. 48-56. Mimeographed; "Acta no. 30 Consejo Ejecutivo del Tratado General de Integración Económica Centroamericana," Guatemala, October 23-November 4, 1967, pp. 48-59. Mimeographed.

40. Sidjanski, *Dimensiones Institucionales de la Integración Latinoamericana*, p. 118.

41. It should be noted that SIECA tried to obtain larger powers when these rules were negotiated, but the governments refused to approve the Secretariat's

proposals. Compare SIECA, "Proyecto de reglamento sobre procedimientos para resolver conflictos" (SIECA/CEC-VIII-O/D.T. 3), Guatemala, November 3, 1967. Mimeographed. With the text approved in SIECA, "Acta vigésimo tercera. Consejo Económico Centroamericano," *La Universidad* (May-August 1968): 301-315.

Chapter 6. Conclusions

1. This notion of "high" and "low" costs is derived from Stanley Hoffmann's notion of "high" and "low" politics as developed in "Obstinate or Obsolete? ," pp. 177-230. See also his "The European Process at Atlantic Crosspurposes," *Journal of Common Market Studies*, 3 (February 1965): 85-101.
2. Haas and Schmitter, "Economics and Differential Patterns," p. 299. Also Schmitter, "La Dinámica de Contradicciones y la Conducción de Crisis en la Integración Centroamericana," *Revista de la Integracón*, no. 5 (November 1969): 146.
3. Schmitter, "La Dinámica de Contradicciones," pp. 140-47.
4. UNCTAD, *Trade Expansion and Economic Integration*, p. 12.
5. Consider, for instance, the existence of six oil refineries in the region; or the fact that "the number of firms that are actually operating is considerably below those classified to receive benefits." IBRD-IDA, "Industry," 4, p. v.
6. Haas and Schmitter, "Economics and Differential Patterns," p. 296.
7. Harry G. Johnson's phrase, *Economic Policies Toward Less Developed Countries* (London: George Allen & Unwin Ltd., 1967), p. 66.

Bibliography

Sources

Banco Central de Costa Rica. "Participación de Costa Rica en la la integración económica centroamericana: Consideraciones Generales" (EE/331/RNM: ASC/mn: ad Agosto 61/50.) San José, August 1961. Mimeographed.

Banco de Guatemala. *Comercio Exterior, Guatemala 1962-1967.* Guatemala, n.d.

_____. "Primera reunión de empresarios guatemaltecos interesados en el Mercado Común Centroamericano, 27-28 de febrero 1967." Guatemala, February, 1967.

Banco Centroamericano de Integracion Economica. "Bases para la formulación de una política regional en materia de inversiones." Tegucigalpa, March 1965. Mimeographed.

_____. "Establecimiento de un sistema de obtención de recursos fiscales para la integración económica centroamericana." Tegucigalpa, May 1967. Mimeographed.

_____. "Investment opportunities in the Central American Common Market." Tegucigalpa, 1967. Mimeographed.

_____. "Estados Financieros." Tegucigalpa, Price Waterhouse and Company, August 9, 1968. Mimeographed.

_____. *Hacia la integración física de Centroamérica*. Tegucigalpa, March 1969.

_____. "Information on the Central American Telecommunications Network and its Financement. Tegucigalpa," n.d. Mimeographed.

_____. "Consideraciones sobre los lineamientos de la política industrial del BCIE y su correspondiente acción crediticia." Tegucigalpa, n.d. Mimeographed.

_____. *Orígenes, Políticas y Gestión*. Tegucigalpa, n.d.

_____. *Memorias de Labores*, 1961-1968. Tegucigalpa.

COCESNA. *Final Minutes and Constitutive Charter, Diplomatic Conference for the Creation of the Central American Corporation for Air Navigation Services*. Tegucigalpa, March 1960.

_____. "Corporación Centroamericana de Servicios de Navegación Aérea." Tegucigalpa, n.d. Mimeographed.

ESAPAC. *Informe del Seminario de Administración del Programa del Convenio Centroamericano de Incentivos Fiscales al Desarrollo Industrial*. (E/CN.12/ CCE/GIF/III/515/No. 1.) San José, February 7, 1962.

_____. *Diagnóstico y macro-análisis administrativos del sector público del Istmo Centroamericano*. (ESAPAC/INV/002/13000/64/RE/3500/65.) San José, October 1965.

_____. *Estudio comparativo de los sistemas tributarios de los países centroamericanos* (ESAPAC/EXT/019/2000/66.) San José, May 1966.

ICAP. "Informe de la investigación sobre el funcionamiento de las aduanas fronterizas centroamericanas." (ICAP/INF/031/150/67.) San José, March 1967. Mimeographed.

――――. *Las empresas públicas del Istmo Centroamericano.* (ICAP/INV/016/66/RE/600/68.) San José, September 1968.

――――. *Importancia de la modernización de la administración pública para el programa de integración del Istmo Centroamericano.* (ICAP/INV/021/500/68.) San José September 1968.

――――. "Recursos humasos, el sector público y su situación actual en Centro America." (ICAP/INV/021/500/68.) San José, January 1968. Mimeographed.

――――. "Necesidades de personal en el sector público de Centroamérica: 1974." (ICAP/INV/024/750/69.) San José, January 1969. Mimeographed.

IBRD-IDA. "Economic Development and Prospects in Central America.", 8 vols. (Report Number WH-170a.), Washington, June 5, 1967. Mimeographed.

    Volume   1: Main Report
                       2: Statistical Appendix to Main Report
                       3: Agriculture
                       4: Industry
                       5: Forestry and Wood Using Industries
                       6: Transportation
                       7: Education
                       8: Water Supply, Sewerage and Public Health

ILO. *La integración económica de América Latina: Problemas de participación y política laboral.* Geneva: ILO, 1968.

ILPES. "Centro América: análisis y provección del estrangulamiento externo en su proceso de desarrollo." Santiago de Chile, August 1966. Mimeographed.

――――. "Posibilidades y modalidades del desarrollo de Centro América (un intento de interpretación sociológica)." Santiago de Chile, January 1967. Mimeographed.

Inter-American Development Bank. "Round Table: The Process of Industrialization in Latin America." Guatemala, Tenth Meeting of the Board of Governors, April 1969. Mimeographed.

――――. "Desarrollo agrícola e integración económica en la América Latina." Washington, April 1969. Mimeographed.

――――. *El BID en Centroamérica.* Washington, December 1968.

Mision Conjunta de Programacion para Centroamerica. "Centroamérica: lineamientos para una política de desarrollo regional." Guatemala, September 1964. Mimeographed.

――――. "Bases para un programa centroamericano de desarrollo industrial." Guatemala, March 1965. Mimeographed.

――――. "Informe sobre el resultado de la coordinación regional de los programas de inversiones públicas." Guatemala, June 1965. Mimeographed.

――――. "Reunión inter-institucional para discutir el documento de la Misión Conjunta: 'Bases para un programa centroamericano de desarrollo industrial' " (ICAITI, SIECA, BCIE, CEPAL, OEA, Misión Conjunta). Guatemala, March 1965. Mimeographed.

———. "Lineamientos de un programa de trabajo para las actividades de planificación en 1966 y 1967." Guatemala, January 1966. Mimeographed.

———. "Presupuesto 1965." Guatemala, n.d. Mimeographed.

———. "Presupuesto para 1966." Guatemala, n.d. Mimeographed.

———. "Borrador de Proyecto de Presupuesto por Programas 1967." Guatemala, n.d. Mimeographed.

Nicaragua, Ministerio de Economía. "Exportaciones-Importaciones de Nicaragua en Centroamérica, Año 1967." Managua, February 29, 1968. Mimeographed.

Organization of American States, Comité de Cooperación OEA-BID-CEPAL. "Informe de labores de la Comisión Preparatoria." Washington, February 13-24, 1961. Mimeographed.

———. "Estudio Económico de América Latina 1962." Washington 1964. Mimeographed.

———. *Sistemas Tributarios de América Latina.*
Guatemala (336-2S-68), Washington 1964.
Costa Rica (336-2S-6845), Washington 1965.
El Salvador (336-2S-6846), Washington 1966.
Honduras (336-2S-6848), Washington 1966.
Nicaragua (336-2S-6849), Washington 1966.

———. Committee of Nine. "Informe sobre los planes nacionales de desarrollo y el proceso de integración económica de Centroamérica." Washington 1966. Mimeographed.

———. Inter-American Economic and Social Council and Inter-American Committee of the Alliance for Progress. "El avance de la integración centroamericana y las necesidades de financiamiento externo." (OEA/SER.H/XIV CIAP 274.) Washington, July 12, 1968. Mimeographed.

"Primera Reunión de Bancos Centrales del Istmo Centroamericano." Tegucigalpa, August, 1952. Mimeographed.

SIECA. "Actas del Consejo Económico Centroamericano 1961-1969." Mimeographed.

———. Actas del Consejo Ejecutivo Centroamericano 1962-1968." Mimeographed.

———. *Inventario de estudios sobre recursos naturales de Centroamérica.* Guatemala, June 1964.

———. "Nota de la Secretaría (sobre inversiones extranjeras)." Guatemala, June 11, 1965. Mimeographed.

———. "Informe sobre los avances del programa de integración económica centroamericana, Diciembre 1964-Enero 1966." (OEA/Ser.H/X.8 CIES 844.) Buenos Aires, February 16, 1966. Mimeographed.

———. "Evaluación de la Secretaría Permanente del Tratado General de Integración Económica Centroamericana (SIECA/CEC-VI-0/120.) Guatemala, April 15, 1966. Mimeographed.

―――. "Lista de mercancías que quedan sujetas a regímenes de excepción al libre comercio al iniciarse el sexto año de vigencia del Tratado General de Integración Centroamericana." (SIECA/SMC/D.I/3-66.) Guatemala, June 1966. Mimeographed.

―――. "Informe preliminar de las industrias clasificadas en los países centroamericanos de conformidad con las leyes nacionales de fomento industrial." (SIECA/DII/128.) Guatemala, June 1966. Mimeographed.

―――. "Proyecto de Reglamento Interno de la Comisión Coordinadora de Mercadeo y Establilización de Precios de Centroamérica." (SIECA/AGROPECUARIA/129) Guatemala, June 11, 1966. Mimeographed.

―――. "Sugerencia de un procedimiento para fijar el monto anual de las importaciones procedentes de terceros países a Centroamérica." (SIECA/AGROPECUARIA/130.) Guatemala, July 11, 1966. Mimeographed.

―――. "Nota de la Secretaría sobre coordinación del desarrollo industrial de Centroamérica." (SIECA/CEC/XIII.E/D.T. 10.) Guatemala, September 14, 1966. Mimeographed.

―――. *5 años de labores en la integración económica centroamericana*. Guatemala, October 12, 1966.

―――. "Formulario para la aplicación de la Resolución No. 26 del Consejo Ejecutivo." (SIECA/CE-XXIII/D.T.4.) San Salvador, October 19, 1966. Mimeographed.

―――. "Problemas del Mercado Común." (SIECA/CE-CIV/D.T.2.) Guatemala, November 8, 1966. Mimeographed.

―――. "Informe sobre los aspectos arancelarios relacionados con la industria textil." (SIECA/CE–XXV/D.T.15.) Guatemala, January 19, 1967. Mimeographed.

―――. "Memorándum sobre reexportación a un tercer país centroamericano de productos originarios de la región incluídos dentro del libre comercio" (SIECA/CE-XXV/D.T.10.) Guatemala, January 16, 1967. Mimeographed.

―――. Proyecto de normas sobre el financiamiento de estudios y dictámenes que se encarguen al ICAITI" (SIECA/CE-XXV/D.T. 24.) Guatemala, January 10, 1967. Mimeographed.

―――. "Memorándum de la sección de desarrollo industrial integrado sobre aspectos de protección arancelaria." (SIECA/CE-XXV/D.T.25.) San José, January 24, 1967. Mimeographed.

―――. "Resoluciones de FECAICA acerca del funcionamiento del programa de integración económica centroamericana." (SIECA/CE-XXV/D.T.26.) San José, January 26, 1967. Mimeographed.

―――. "La comercialización de los principales productos agropecuarios en el proceso de integración centroamericana." (SIECA/AGROPECUARIA/ Informativo.) Guatemala, March 16, 1967. Mimeographed.

_____. "El algodón en Centroamérica." (SIECA/Agropecuaria/Informativo.) Guatemala, March 16, 1967. Mimeographed.

_____. "Informe sobre los avances del programa de integración económica centroamericana (Febrero 1966-Mayo 1967)." Viña del Mar, June 15-26, 1967. Mimeographed.

_____. "Resoluciones del Consejo Económico." Guatemala, June 30, 1967. Mimeographed.

_____. "Resoluciones del Consejo Ejecutivo." Guatemala, June 30, 1967. Mimeographed.

_____. "Propuesta de organización institucional para la política comercial centroamericana." (SIECA/CEC-VII-O/D.T.5.) Guatemala, July 20, 1967. Mimeographed.

_____. "Proyecto de Presupuesto por Programas de la Secretaría Permanente del Tratado General de Integración Económica Centroamericana 1968." (SIECA/CEC-VIII-O/D.T.2.) Guatemala, October 5, 1967. Mimeographed.

_____. "Inflexibilidad del arancel uniforme y la integración Centroamericana" (SIECA/CEC/CMCA/MH-1-D.T.3.) Guatemala, October 14, 1967. Mimeographed.

_____. "El financiamiento de las instituciones regionales de la integración económica centroamericana." (SIECA/CEC/CMCA/MH-1-D.T.3.) Guatemala, October 31, 1967. Mimeographed.

_____. "Informe de actividades de la Secretaría Permanente del Tratado General de Integración Económica Centroamericana, Enero-Septiembre 1967." (SIECA/CEC-VIII-O/D.I.1.) Guatemala, November 7, 1967. Mimeographed.

_____. "Primer borrador de proyecto de protocolo para manejar en forma flexible el arancel uniforme centroamericano." (SIECA/CEC/CMCA/MH-1-D.T.7.) Managua, November 14, 1967. Mimeographed.

_____. "Resumen y estado actual de los tratados de integración económica centroamericana," Guatemala, March 14, 1969. Mimeographed.

_____. *Convenios centroamericanos de integración económica*. 3 vols. Guatemala, 1963-1964.

_____. "Informe sobre los avances del programa de integración económica centroamericana, junio 1967-mayo 1969." (SIECA/69/S.G.52.) June 9, 1969. Mimeographed.

_____. -CEPAL. "Los problemas de la política industrial centroamericana." (SIECA/CEC-III/Prov. 30.) January 14, 1964. Mimeographed.

_____. -Secretaría Ejecutiva del Consejo Monetario Centroamericano. "El problema de la balanza de pagos y la integración económica centroamericana." (SIECA/CEC/CMCA/MH-1-D.T.2) Guatemala, October 5, 1967. Mimeographed.

"Tercera Reunión de Bancos Centrales Centroamericanos." Guatemala, February 21-26, 1955. Mimeographed.

UNCTAD. *Trade Expansion and Economic Integration among Developing Countries*. (Sales No.:67.II.D.20.) New York, 1967.

UN-ECLA. "Actas y Documentos de las Reuniones del Comité de Cooperación Económica," 1952-1966.

———. "Actas y Documentos de las Reuniones de los organismos del Comité de Cooperación Económica," 1954-1968. Mimeographed.

———. *El Transporte en el Istmo Centroamericano*. (E/CN.12/356.) México, September 1953. (Sales No.: 1953.VIII.2.)

———. *La política tributaria y el desarrollo económico de Centro América*. (Sales No.: 1957.II.69.) September 1956.

———. "Informe de la Primera Reunión de Inversionistas Centroamericanos." (E/CN.12/CCE/206.) San Salvador, October 27-31, 1959. Mimeographed.

———. "El programa de integración económica de Centro América y el Tratado de Asociación Económica suscrito por El Salvador, Guatemala y Honduras." (E/CN.12/CCE/212.) May 5, 1960. Mimeographed.

———. "Proyecto de Tratado General de Integración Económica Centroamericana, Nota de la Secretaría" (E/CN.12/CCE/S.C.1/58.) October 14, 1960. Mimeographed.

———. "Posibilidades de Desarrollo Industrial Integrado en Centro América. Volumen I." (Sales No.: 63.II.G.10.) November 1963. "Volumen II." (E/CN. 12/CCE/323/Rev. 1.) August 12, 1965. Mimeographed.

———. "El sector externo y el desarrollo económico de Centro América 1950-1962." Nota Informativa de la Secretaría. (E/CN.12/CCE/Sc.1/I R.EXT/DI 2.) January 1964. Mimeographed.

———. "Carreteras, Puertos y Ferrocarriles de Centroamérica." (E/CN.12/CCE/ 324.) August 10, 1965. Mimeographed.

———. "La institucionalización regional de la planificación en Centroamérica." (Comité Asesor/VI/2/Rev. 1.) August 25, 1965. Mimeographed.

———. *Evaluación de la Integración Centroamericana*. (Sales No.: 66.II.G.9.) New York, 1966.

———. "La productividad industrial, el costo de la mano de obra y el costo de producción en el Istmo Centroamericano." (E/CN.12/CCE/335/Rev. 1.) April 4, 1966. Mimeographed.

———. "Guía Básica de la Comisión y de su Secretaría." July 1966. Mimeographed.

———. "Indices de Resoluciones de la CEPAL 1948-1966." August 1966. Mimeographed.

———. "Los países de menor desarrollo relativo y la integración latinoamericana." (E/CN.12/774.) April 8, 1967. Mimeographed.

———. "Informe de la reunión sobre los problemas de la integración regional de los países de menor desarrollo relativo." (E/CN.12/798.) February 13, 1968. Mimeographed.

———. "Consideraciones sobre la adopción de una política comercial externa común para los países centroamericanos." (E/CN.12/CCE/S.C.1/97.) November 1967. Mimeographed.

_____. "Informe de la Secretaría sobre el Programa de Integración Económica Centroamericana, Mayo 1967-Abril 1968." (CEPAL/MEX/68/4.) February 16, 1968. Mimeographed.

_____. "Breve reseña de las actividades de la subsede de la CEPAL en México desde su creación en 1951 hasta mayo de 1968." (CEPAL/MEX/68/11.) April 1968. Mimeographed.

_____. "Resoluciones del Comité de Cooperación Económica del Istmo Centroamericano," Volume 1: "Indice Correlativo Clasificado"; Volume 2: "Textos." (E/CN.12/CCE/258.) July 1968. Mimeographed.

_____. "Situación y tendencias demográficas recientes en Centroamérica." (E/CN.12/CCE/356.) May 15, 1968. Mimeographed.

_____. "Distribución de la población en el Istmo Centroamericano." (E/CN.12/CCE/357.) August 1, 1968. Mimeographed.

_____. "Aspects of the Interrelations between the Trends of Economic Development and Human Resources in México, Central America and Panamá." (CEPAL/MEX/68/14.) August 12, 1968. Mimeographed.

_____. "Subsede de la CEPAL en México: Programa de Trabajo para 1970-1971." (CEPAL/MEX/68/15/Rev. 1.) November 13, 1968. Mimeographed.

UN-ECOSQC. *Official Records*, 2nd year, 5th session; from the 85th meeting (July 19, 1947) to the 121st meeting (August 16, 1947). Lake Success, New York, 1948.

US-AID-ROCAP. "The Intra-Regional Trade of Central America—Its Magnitude, Composition, and Changing Patterns." Guatemala, July 16, 1965. Mimeographed.

_____. "Central America: Public External Debt and Debt Service." Guatemala, June 6, 1966. Mimeographed.

_____. "The 'foreign travel' component in the Central American Balance of Payments." Guatemala, n.d. Mimeographed.

_____. *Unión Monetaria Centroamericana*. México, December 28, 1964.

_____. "The Central American Common Market." Guatemala, September 1962. Mimeographed.

_____. "The Operations of the Central American Common Market." Guatemala, August 19, 1966. Mimeographed.

US Congress, Joint Economic Committee, Sub-Committee on Inter-American Relations, *Hearings on Private Investment in Latin America*. 88th Cong., 2nd Sess., 1964.

_____. *Hearings on Latin American Development and Western Hemisphere Trade*. 89th Cong., 1st Sess., 1965.

_____. House of Representatives, Committee on Foreign Affairs, Sub-Committee on Inter-American Affairs, *Central America: Some Observations on its Common Market, Binational Centers, and Housing Programs*. (Report by Hon. Roy H. McVicker) 89th Cong., 2nd Sess., Washington: Government Printing Office, 1966.

US Senate. *United States and Latin American Policies Affecting Their Economic Relations.* Washington: Government Printing Office, 1960. A study prepared at the request of the Sub-Committee on American Republics Affairs of the Committee on Foreign Relations by the National Planning Association.

TSC Consortium. "Central American Transportation Study, Summary Report." Guatemala, November 1965. Mimeographed.

Works

Banco Centroamericano de Integración Económica. *Un Mercado de Capitales Centroamericano: Dos Estudios.* México: CEMLA, 1967.

Calderón Gonzalez, Juan Arnoldo. *Endeudamiento Externo de Guatemala.* Guatemala: Universidad de San Carlos, 1969.

Castillo, Carlos Manuel. *Growth and Integration in Central America.* New York: Frederik A. Praeger, 1966.

Cochrane, James D. *The Politics of Regional Integration: The Central American Case.* The Hague: Martinus Nijhoff, 1969.

Colegio de Abogados de Guatemala. *Aspectos jurídicos e institucionales de la integración centroamericana.* Guatemala, 1967.

Committee for Economic Development. *Economic Development of Central America.* Washington, 1964.

Consejo Monetario Centroamericano. *Hacia la Unión Monetaria Centroamericana.* San José, 1968.

Camara de Compensación Centroamericana. *Tres Años de Cooperación Multilateral.* Tegucigalpa, n.d.

Cruz, Ernesto. *Derecho, Desarrollo e Integración Regional.* San Salvador: ODECA, 1967.

Diaz Reyes, José Vicente. "Las Industrias de Integración en Centroamérica." Tegucigalpa: Universidad Nacional Autónoma de Honduras, 1966. Mimeographed.

Facio, Rodrigo. *La Federación de Centroamérica: Sus Antecedentes, Su Vida y Su Disolución.* San José: ESAPAC, 1965.

Gonzalez Dubón, Cristina Idalia. "Bibliografía Analítica sobre la Integración Económica Centroamericana." Guatemala: Universidad de San Carlos, 1969. Mimeographed.

Graciarena, Jorge. *Poder y clases sociales en el desarrollo de América Latina.* Buenos Aires: Paidós, 1967.

Haas, Ernst B. *Beyond the Nation-State: Functionalism and International Organization.* Stanford: Stanford University Press, 1964.

Hansen, Roger D. *Central America: Regional Integration and Economic Development.* Washington: National Planning Association, 1967.

Herrera Zúñiga, René. *Estudio acerca de los aspectos fundamantales del Tratado General de Integración Centroamericana.* México: Universidad Nacional Autónoma de Guadalajara, 1967.
Hirschman, Albert O. *Journeys Toward Progress.* New York: The Twentieth Century Fund, 1963.
Hoffmann, Stanley. *Gulliver's Troubles or the Setting of American Foreign Policy.* New York, McGraw-Hill Book Company for the Council on Foreign Relations, 1968.
Instituto Interamericano de Estudios Jurídicos Internacionales. *El Sistema Interamericano.* Madrid: Ediciones Cultura Hispánica, 1966.
──── . *Derecho Comunitario Centroamericano.* San José: Trejos Hermanos, 1968.
Jimenez Lazcano, Mauro. *Integración Económica e Imperialismo.* México: Editorial Nuetro Tiempo, 1968.
Johnson, Harry G. *Economic Policies Toward Less Developed Countries.* London: George Allen & Unwin Ltd., 1967.
Lizano, Eduardo. "La integración económica centroamericana y la distribución del ingreso." San José, 1968. Mimeographed.
McCamant, John F. *Development Assistance in Central America.* New York: Frederick A. Praeger, 1968.
Organizacion de Estados Centroamericanos. *Reuniones y Conferencias de Ministros de Relaciones Exteriores de Centroamérica 1951-1967.* San Salvador: ODECA, 1967.
Schlesinger Jr., Arthur M. *A Thousand Days.* London: Mayflower Books, 1965.
Sidjanski, Dusan. *Dimensiones Institucionales de la Integración Latinoamericana.* Buenos Aires: INTAL, 1967.
Villagrán Kramer, Francisco. *Integración Económica Centroamericana: Aspectos Sociales y Políticos.* Guatemala: Universidad de San Carlos, 1967.
Zeledón, Marco Tulio. *La ODECA, sus antecedentes históricos y su aporte al Derecho Internacional Público.* San José: Colegio de Abogados de Costa Rica, 1966.
Zinner, Paul E., ed. *Documents on American Foreign Relations, 1959.* New York: Harper & Brothers for the Council on Foreign Relations, 1960.

Articles

Anderson, Charles W. "Politics and Development Policy in Central America." In *Latin American Politics*, edited by Robert D. Tomasek. New York: Anchor Books, 1966, pp. 544-64.
Barrera, Mario and Haas, Ernst B. "The Operationalization of Some Variables Related to Regional Integration: A Research Note." *International Organization* 23: (Winter 1969) 150-60.

Caballeros, Jorge Lucas. "Influencia del mercado común centroamericano en el desarrollo económico de Guatemala." *Universidad de San Carlos*, no. 71 (July-September 1967), pp. 139-46.
Castillo, Carlos Manuel. "Estado del Programa de Integración Económica Centroamericana." *Universidad de San Carlos*, no. 71 (July-December 1967), pp. 5-28.
──── . "Centroamérica y la integración económica latinoamericana." San Pedro Sula, June 7, 1967, pp. 12. Mimeographed.
Castrillo Zeledón, Mario. "La zona de libre comercio en Centroamérica." *La Universidad*, nos. 3-4 (May-August 1968), pp. 11-44.
──── . "La integración económica centroamericana y los sindicatos de trabajadores." *La Universidad*, nos. 3-4 (May-August 1968), pp. 233-44.
Cochrane, James D. "US Attitudes toward Central American Integration." *Inter-American Economic Affairs*, vol. 18, no. 2 (Autumn 1964), pp. 73-91.
──── . "Central American Economic Integration and the 'Integrated Industries' Scheme." *Inter-American Economic Affairs*, vol. 19, no. 2 (Autumn 1965), pp. 63-74.
Collart Valle, Antonio. "Política comercial, tratados de comercio y unificación arancelaria como factores en el proceso de integración económica centroamericana." In *Integración Económica de Centroamérica*, edited by Jorge Luis Arriola. San Salvador: ODECA, 1959, pp. 307-26.
Dalponte, Mario. "La mecánica de la integración económica como medio complementario de desarrollo: observaciones críticas en torno al programa centroamericano." *La Universidad*, nos. 3-4 (May-August 1968), pp. 45-54.
Delgado, Pedro Abelardo. "Origines, Structure et Fonctionement du Marché Commun Centre-Américain." *Tiers Monde*, tome 6, no. 23 (July-September 1965), pp. 643-62.
Deutsch, Karl; Burrel, Sidney A.; Kann, Robert A.; Lee Jr., Maurice; Lichterman, Martin; Lindgren, Raymond E.; Loewenhenheim, Francis L.; and Van Wagenen, Richard W., "Political Community in the North Atlantic Area." In *International Political Communities: An Anthology*. New York: Anchor Books, 1966, pp. 1-91.
Dick, Philip J. "The role of the mass media in Central American Development." Studies of Entrepreneurial Behavior in Central America, Project Director: Calvin P. Blair, 1968, p. 24. Mimeographed.
Duarte Fuentes, Gustavo. "Movimientos de capital en Centroamérica: Alguna experiencia en la Cámara de Compensación Centroamericana y sus perspectivas." In *Compendio de Estudios Técnicos presentados al II Congreso Centroamericano de Economistas, Contadores Públicos y Auditores*. San Salvador: INSAFI, 1965, pp. 405-418.
Eckenstein, Christopher. "La integración económica en Centroamérica: Un ejemplo de cooperación exitosa entre países en vías de desarrollo." "Neue Zürher Zeitung," November 29, December 1,4,5, 1964. Mimeographed Spanish version.

———. "Integración Regional: más realismo." *Ceres.* vol. 1, no. 6 (November-December 1968), pp. 22-25.

Fajardo Maldonado, Arturo. "Soberanía y Derecho Internacional en el proceso de integración económica centroamericana." *La Universidad*, nos. 3-4 (May-August 1968), pp. 147-76.

Fonseca, Gautama. Las Fuentes del Derecho Común Centroamericano." *La Universidad*, nos. 3-4 (May-August 1968), pp. 127-46.

Gibson, Cyrus Frank. "Regional Foreign Aid in Central America: The Implications." Studies of Entrepreneurial Behavior in Central America, Project Director: Calvin P. Blair, 1968, p. 24. Mimeographed.

Glower Valdivieso, Rafael. "La economía salvadoreña a la luz de la economía centroamericana." *La Universidad*, nos. 3-4 (May-August 1968), pp. 55-62.

Gonzalez del Valle, Jorge. "Sistemas de Pagos Inter-centroamericanos." In *Integración de América Latina, Experiencias y Perspectivas*, edited by Miguel S. Wionczek. México: Fondo de Cultura Económica, 1964, pp. 313-26.

———. "Monetary Integration in Central America: Achievements and Expectations." *Journal of Common Market Studies*, vol. 5, no. 1 (September 1966), pp. 13-25.

Gregg, Robert W. "The UN Regional Economic Commissions and Integration in the Underdeveloped Regions." In *International Regionalism*, edited by J.S. Nye. Boston: Little Brown & Co., 1968, pp. 304-332.

Groom, A.J.R. "Participation, Functionalism and Systems at the World Level." Carnegie Endowment for International Peace. Paper submitted to the Conference on Functionalism and the Changing Political System. Bellagio, November 20-24, 1969, pp. 22. Mimeographed.

Guillen V., José. "Integración Agrícola de Centroamérica." *Universidad de San Carlos*, no. 71 (July-December 1967), pp. 171-84.

Haas, Ernst B. "International Integration: The European and the Universal Process." In *International Political Communities: An Anthology*. New York: Anchor Books, 1966, pp. 93-129.

———. "The Uniting of Europe and the Uniting of Latin America." *Journal of Common Market Studies*, vol. 5, no. 4 (June 1967), pp. 315-43.

———. "Future Worlds and Present International Organizations: Some Dilemmas." *Bulletin International Institute of Labour Studies*, no. 6 (June 1969), pp. 4-53.

Haas, Ernst B. and Schmitter, Philippe. "Economics and Differential Patterns of Political Integration: Projections about Unity in Latin America." In *International Political Communities: An Anthology*. New York: Anchor Books, 1966, pp. 259-299.

Hernández Segura, Roberto. "La unión monetaria centroamericana en el proceso de integración económica." In *Compendio de Estudios Técnicos presentados al II Congreso Centroamericano de Economistas, Contadores Públicos y Auditores.* San Salvador: INSAFI, 1965, pp. 391-402.

Hirschman, Albert O. "La economía política de la industrialización a través de la sustitución de importaciones en América Latina." *El Trimestre Económico*, vol. 35 (4), no. 140, (October-December 1968), pp. 625-58.

Hoffmann, Stanley. "Obstinate or Obsolete? The Fate of the Nation-State and the Case of Western Europe." In *International Regionalism*, edited by J.S. Nye. Boston: Little Brown & Co., 1968, pp. 177-230.

―――. "The European Process at Atlantic Crosspurposes." *Journal of Common Market Studies*, vol. 3, no. 2 (February 1965), pp. 85-101.

Jiménez Castro, Wilburg. "Necesidad de la formación de ejecutivos y su importancia en el desarrollo económico." *Compendio de los Estudios Técnicos Presentados al II Congreso Centroamericano de Economistas, Contadores Públicos y Auditores.* San Salvador: INSAFI, 1965, pp. 35-38.

Kaiser, Karl. "The Interaction of Regional Subsystems: Some Preliminary Notes on Recurrent Patterns and the Role of Superpowers," *World Politics*, vol. 21, no. 1, pp. 84-107.

Lagos, Gustavo. "L'intégration de l'Amérique Latine et son influence sur le système international." *Tiers Monde*, tome 6, no. 23 (Juillet-Septembre 1965), pp. 743-55.

―――. "The Political Role of Regional Economic Organizations in Latin America." *Journal of Common Market Studies*, vol. 6, no. 4, pp. 291-309.

Lara, Gilberto. "Breve análisis del concepto de 'desarrollo equilibrado' en el ámbito de la integración económica centroamericana." *Compendio de los Estudios Técnicos presentados al II Congreso Centroamericano de Economistas, Contadores Públicos y Audiores.* San Salvador: INSAFI, 1965, pp. 367-78.

Levine, Meldon E. "El sector privado y el Mercado Común: Reacciones de la iniciativa privada de Honduras, Nicaragua y El Salvador con respecto al Mercado Común Centroamericano." Guatemala: Woodrow Wilson School of Public and International Affairs and Instituto Nacional de Administración para el Desarrollo (INAD), August 1968. Mimeographed Spanish version.

Lindberg, Leon. "The European Community as a Political System: Notes Toward the Construction of a Model." *Journal of Common Market Studies*, vol. 5, no. 4 (June 1967), pp. 84-107.

―――. "Decision Making and Integration in the European Community." In *International Political Communities: An Anthology*. New York: Anchor Books, 1966, pp. 199-231.

Lizano, Eduardo. "La crisis del proceso de integración centroamericana." San José: Universidad de Costa Rica, 1965, pp. 35. Mimeographed.

Martén Chavarría, Alberto. "Canalización del ahorro particular para el desarrollo de un mercado centroamericano de valores," in *Integración Económica de Centroamérica*, edited by Jorge Luis Arriola. San Salvador: ODECA, 1959, pp. 307-326.

Mills, Joseph C. "Problemas de la Industrialización en la América Central." In *Integración de América Latina, Experiencias y Perspectivas*, edited by Miguel S. Wionczek. México: Fondo de Cultura Económica, 1964, pp. 292-305.

Molina Calderon, José. "El peso centroamericano: situación y perspectivas." *Universidad de San Carlos*, no. 71, (July-December 1967), pp. 29-66.

Monteforte Toledo, Mario. "Los intelectuales y la integración centroamericana." *Revista Mexicana de Sociología*, vol. 29, no. 4 (October-December 1967), pp. 831-52.

_____."La integración económica y el panorama político de Centroamérica." *La Universidad*, nos. 3-4 (May-August 1968), pp. 63-76.

Moscarella, Joseph. "La Integración Económica Centroamericana." In *Integración de América Latina, Experiencias y Perspectivas*, edited by Miguel S. Wionczek. México: Fondo de Cultura Económica, 1964, pp. 273-91.

Noriega Morales, Manuel. "La localización industrial y la integración económica centroamericana." In *Integración Económica de Centroamérica*, edited by Jorge Luis Arriola. San Salvador: ODECA, 1959, pp. 173-228.

Nye, Joseph. "Patterns and Catalysts in Regional Integration." *International Organization*, vol. 19, no. 4 (Autumn 1965), pp. 333-49.

_____. "Central American Regional Integration." *International Conciliation*, no. 562 (March 1967).

_____. "Comparative Regional Integration: Concept and Measurement." *International Organization*, vol. 22, no. 4 (Autumn 1968), pp. 855-80.

_____. United States Policy Toward Regional Organization." *International Organization*, vol. 23, no. 3 (Summer 1969), pp. 719-40.

Nugent, Jeffrey B. "La estructura arancelaria y el costo de protección en América Central." *El Trimestre Económico*, vol. 35(4), no. 140 (October-December 1968), pp. 751-66.

Ordoñez Fernandez, Hugo. "El Banco Centroamericano en el proceso de integración económica." *Universidad de San Carlos*, no. 71 (July-December 1967), pp. 75-94.

Padelford, N.J. "Cooperation in the Central American Region: The Organization of Central American States." *International Organization*, vol. 11, no. 1 (Winter 1957), pp. 41-54.

Palacios, J. Antonio. "Los emprearios en la integración." *Universidad de San Carlos*, no. 71 (July-December 1967), pp. 109-116.

Pentland, C. "Functionalism and Neofunctionalism: Some Comments." Carnegie Endowment for International Peace. Paper submitted to the Conference on Functionalism and the Changing Political System. Bellagio, November 20-24, 1969, pp. 12. Mimeographed.

Von Potobsky, Geraldo. "La participación de las asociaciones profesionales en el proceso de planificación de los países de América Latina." *Revista Internacional del Trabajo*, vol. 75, no. 6 (June 1967), pp. 3-24.

Prebisch, Raúl. "The Economic Development of Latin America and Some of its Principal Problems." *Economic Bulletin for Latin America*, vol. 7, no. 1 (February 1962), pp. 1-22.

Ramirez, Marco Antonio. "Movilidad de mano de obra e integración económica centroamericana." In *Integración Económica de Centroamérica*, edited by Jorge Luis Arriola. San Salvador: ODECA, 1959, pp. 251-58.

──────. "Comercio de Alimentos en el Mercado Común Centroamericano y en el de Panamá." *Universidad de San Carlos*, no. 68 (January-June 1966), pp. 39-83.

Rosenthal, Gert. "Consideraciones acerca del 'desarrollo equilibrado.' " *Universidad de San Carlos*, no. 71 (July-December 1967), pp. 95-102.

──────. "Algunas consideraciones acerca del 'Desarrollo Equilibrado' en el desenvolvimiento de la Integración Económica Centroamericana." Guatemala: Consejo Nacional de Planificación Económica, Memorandum 3-66, January 17, 1966. Mimeographed.

Schmitter, Philippe. "La Dinámica de Controadicciones y la Conducción de Crisis en la Integración Centroamericana." *Revista de la Integración*, no. 5 (November 1969), pp. 87-151.

──────. "Three Neo-Functional Hypotheses About International Integration." *International Organization*, vol. 23, no. 4 (Winter 1969), pp. 161-66.

──────. "Further Notes on Operationalizing Some Variables Related to Regional Integration." *International Organization*, vol. 23, no. 2 (Spring 1969), pp. 327-36.

Schmitter, Philippe and Haas, Ernst B. "Méjico y la Integración Económica Latinoamericana." *Desarrollo Económico*, vol. 4, no. 14-15 (July-December 1964), pp. 111-69.

Segal, Aron. "The Integration of Developing Countries: Some Thoughts on East Africa and Central America." *Journal of Common Market Studies*, vol. 5, no. 3 (March 1967), pp. 252-82.

Sierra Franco, Raúl. "La industrialización dentro del programa de integración económica de Centro América." *Economía*, año 1, no. 1 (January-March 1962), pp. 31-35.

Siotis, Jean. "The Secretariat of the United Nations Economic Commission for Europe and European Economic Integration: The First Ten Years." *International Organization*, vol. 19, no. 2 (Spring 1965), pp. 177-202.

Sol Castellanos, Jorge. "La integración económica de Centroamérica y los programas nacionales de desarrollo económico." In *Integración Económica de Centroamérica*, edited by Jorge Luis Arriola. San Salvador: ODECA, 1959, pp. 37-76.

Staley, S. "Costa Rica and the Central American Common Market." *Economia Internazionale*, vol. 15, no. 1 (February 1962), pp. 117-30.

Taylor, Paul. "The Functionalist Approach to Integration." Carnegie Endowment for International Peace, paper submitted to the Conference on Functionalism and the Changing Political System. Bellagio, November 20-24, 1969, pp. 11. Mimeographed.

Teubal, Miguel. "The Failure of Latin America's Economic Integration." In

Latin America: Reform or Revolution? edited by James Petras and Maurice Zeitlin. Greenwich, Conn.: Fawcet Publications Inc., 1968, pp. 120-44.

Tratado de Intercambio Preferencial y de Libre Comercio entre las Repúblicas de Panamá, Nicaragua y Costa Rica. *Revista de Economía,* año 5, no. 1 (August 1963), pp. 1-5.

Triffin, Robert and Cooper, Richard N. "Propuesta para crear un Fondo Centroamericano de Estabilización." San José: Consejo Monetario Centroamericano, June 1968. Mimeographed Spanish version.

UN-ECLA "El programa de integración económica de Centroamérica." *Boletín Económico de América Latina,* vol. 4, no. 2 (October 1959), pp. 3-18.

———. "Central America's post-war exports to the United States." *Economic Bulletin for Latin America,* vol. 5, no. 2 (October 1960), pp. 24-56.

———. "Estado General y Perspectivas del Programa de Integración Económica del Istmo Centroamericano." *Boletín Económico de América Latina,* vol. 8, no. 1 (March 1963), pp. 9-25.

———. "The Central American Common Market for Agricultural Commodities." *Economic Bulletin for Latin America,* vol. 10, no. 1 (March 1965), pp. 23-47.

———. "Vigésimo Aniversario de la CEPAL." *Boletín Económico de América Latina,* vol. 13, no. 2 (November 1968), pp. 139-40.

Vasquez, Alexander. "Transportes y Comunicaciones en la Integración Económica." *Economía,* año 4, no. 9 (April-June 1965), pp. 33-42.

Velasquez, Roberto. "La coexistencia de leyes nacionales de protección industrial y los convenios de alcance regional." *Universidad de San Carlos,* no. 71 (December 1967), pp. 127-38.

Villagrán Kramer, Francisco. "Adecuación de la legislación laboral salvadoreña a los requerimientos del mercado común centroamericano." *La Universidad,* nos. 3-4 (May-August 1968), pp. 221-32.

Walter, Ingo and Vitzthum, Hans C. "The Central American Common Market: A Case Study on Economic Integration in Developing Regions." *The Bulletin,* no. 44 (May 1967).

Wionczek, Miguel. "Condiciones de una integración viable." In *Integración de América Latina, Experiencias y Perspectivas,* edited by Miguel S. Wionczek. México: Fondo de Cultura Económica, 1964, pp. xvii-xxxi.

———. "Integración económica y distribución regional de las actividades industriales (estudio comparativo de las experiencias de Centroamérica y el Africa Oriental)." *El Trimestre Económico,* vol. 33(3), no. 131 (July-September 1966), pp. 469-502.

———. "Mercados Comunes Latinoamericanos: ¿Hacia dónde van?" *Ceres,* vol. 1, no. 6 (November-December 1968), pp. 32-37.

———. "Latin American Integration and United States Economic Policies." In *International Organization in the Western Hemisphere,* edited by Robert W. Gregg. New York: Syracuse University Press, 1968, pp. 91-156.

Periodicals

Carta Informativa del Banco Centroamericano de Integración Económica. Tegucigalpa.
Carta Informativa de la SIECA. Guatemala.
Department of State Bulletin. Washington, D.C.

Index

Agency for International Development (AID), 37, 50, 63, 74
Agriculture, 43-44, 64, 86
  effect of the program on, 55-56
  importance of, 2
  sectors, 4
Alcoholic beverages, 43
Alliance for Progress, 35, 38, 39, 62
Antitrust laws, 44
Assembly goods, 42, 43, 45
Attitudes: toward ECLA program, 24
  toward free trade, 44-45
  toward integration, 10
  of the United States, 27-29, 35
  toward the United States, 27-28
Autonomy, institutional, 73-82

Balance of payments, 58-59
Balanced development, 36-37, 47, 77-80
Banana companies, 2, 3, 8, 11

Capital goods, protection of, 46
Caribbean Legion, 6
Cattle, 55
Central American Bank of Economic Integration (CABEI), 35, 38, 50-51, 54, 60, 62, 69, 70-71
Central American Corporation of Air Navigation Services (COCESNA), 52, 71
Central American Institute of Industrial Technology and Research (ICAITI), 22, 62-63, 67, 71, 74
Central American Institute of Public Administration (ICAP), 63, 67, 71, 74
Central American Monetary Council, 71
Central American School of Public Administration (ESAPAC), 22, 63, 67
Central banks, 57, 58, 59, 71
Cereals, 55
Charter of San Salvador, 16-17
Changes: policy, 28-29, 35
  political, 3-4, 14
Cheese, 43
Cigarettes, 43
Cinema seats, number of, 11
Clearing house, regional, 57
Coffee, 43, 55
Commercial policy, 19-20, 41, 72-73

Committee of Economic Cooperation, 17, 22, 34, 35, 50, 54, 67, 68, 69, 70, 79
Common external tariff, 23-24, 32, 35, 42, 45-47
Common market, 34, 41
Communications, 3, 8, 11, 52, 53-54
Communism, 17
Competition among governments, 30, 46-47, 48, 49, 61, 65, 80
Conditions (regional): favoring integration, 11
  indicators of, 1
Conflicts (regional), 6-7
  resolution of, 80-84
Construction materials, protection of, 46
Consumption goods, protection of, 46
Convertibility, currency, 31, 32, 57
Cooperation: institutional, 71
  monetary, 56-59
  regional, 17, 52
Coordinating commission of marketing and price stabilization, 55-56
Coordination of program activities, 50, 55-56, 67, 69, 71-72
  lack of, 52, 58, 69, 70, 71-72, 73
Costs (program), 20-22, 24
  avoidance of, 83-84
  political, 85
Cotton, 43, 55, 58
Credit, 57
Cuban revolution, 28, 35-36, 39
Customs union, 21, 23, 29, 38, 42, 44

"Declaration of Central America," 38
Delegations, members of, 15
Demand: expanding, 15
  increases in, 49
Dictatorships, 4
Dillon, Douglas, 29

Echandi, Mario, 33
Economic Commission for Latin America (ECLA), 13-25, 62
  aims and objectives, 13, 29, 47
  hegemony, 21-25, 67
  Mexico City Office, 16, 18, 21, 22, 24, 33, 34, 67, 69, 79
  proposals by, 13, 14, 15, 18-19
  reaction to Tripartite Treaty, 33-34

123

reduced participation, 24-25, 69
role, 5, 69
strategy, 15-21
studies by, 21, 23
US opposition to, 28-29, 39
   See also Committee of Economic Cooperation
Economic Council, 17, 18, 62, 68, 69, 70, 72, 78, 82
Economic integration, separation from political integration, 15-18
Education, 63
Electric power, 53
Elites, 4, 11
  values of, 3
Entrepreneur, 3, 76
ESAPAC, 22, 63, 67
Executive Council, 67, 68, 70, 72, 80, 81
Exiles, role of, 6, 35
Export, 2

Farms, 4
Financing: of CABEI, 60
  of ESAPAC, 22
  of ICAP, 74
  of ICAITI, 22, 63, 74
  of Joint Planning Mission, 62
  of ODECA, 74
  regional contributions toward, 22, 39, 50, 52, 56, 57, 60, 63, 74
  of roads, 50-51
  of telecommunications, 54
    See also Foreign aid
Fiscal incentives, agreement for the unification of, 48
Foodstuffs, share of trade, 43
Foreign aid, 38, 51, 52, 56, 59, 60, 61, 74, 84, 86
  US, 8, 28, 32, 35, 37, 38, 39, 50, 60
Free trade, 7, 32, 34, 42-45
  conflicts created by, 80-82
  ECLA aim of, 29
  exclusions from, 24, 32, 42-44, 45, 55
  list of products for, 23
  obstacles to, 44, 80
  treaties, 19, 23-24, 30, 67, *see also* Protocols
Free trade area, 21, 29, 41
Fund: Central American, 38, 39, 50
  Special, UN, 74

General Agreement on Tariffs and Trade (GATT), 29-30
General Treaty of Central American Economic Integration, 34-35, 38, 39, 55, 62, 68, 69, 83
González del Valle, 58
Government: legitimacy of, 5, 6, 7
  responsibilities of, 4-5, 14
Gradual integration, 19-20
Grains, regulation of, 55-56
Great depression, 4
Gross domestic product (GDP), 2
Growth: rates of, 2
    See also Balanced development
Guarantee fund, 57

Hoffman, Stanley, 40
Homogeneity, 3
Honduran Association of Industrialists, 78

ICAITI, 22, 62-63, 67, 71, 74
ICAP, 63, 67, 71, 74
IDB, 50, 54, 60, 62
Immigrants, 10
Import substitution as a solution, 13, 14, 19, 23, 86
Incentives, fiscal, agreement on, 78-79
Income: per capita, 64
  government, 3
Indicators of regional conditions, 1
Industrialization, 19-21, 65
  need for, 14
  post-war, 2
  problems of, 47-49
  reciprocal, 20
Industries, 2, 20, 49
  location of, 47
  make-up, 65
  planned distribution of, 23
Institutions (regional), 67-83
  autonomy of, 73-82
  created 1951-1959, 67
  economic, 72
  financing for, 74-75
  isolation of, 71
  role of, 76, 80, 82
Inter-American Development Bank (IDB), 50, 54, 60, 62
Inter-American Highway, 8, 10, 50
Inter-American Treaty of Reciprocal Assistance, 6
International Civil Aviation Organization (ICAO), 52
International Court of Justice, 6
International Monetary Fund (IMF), 28
International Telecommunications Union (ITU), 54
Investment: competition for, 46
  incentives for, 48

plan for, 19
private foreign, 60-61

Joint Planning Mission for Central America, 62, 71, 78

Kennedy, President, 35, 38, 39, 50, 58

Labor Unions, 76
LAFTA, 24
Land: ownership, 4
   tenure, 3-4
Latin American Free Trade Association (LAFTA), 24
Legitimacy of government, concept of, 5, 6, 7
Limited integration, 19
Literacy, 11
Loans, 50, 60, 61

Mann, Thomas, 31, 33, 39
Manufactures, share of trade, 43
Manufacturing, 65
   growth of, 48-49
   importance of, 2
Market(s): creation of, 86
   expansion of, 64
   size of, 14
Matches, 43
Migration, 10-11
Milk, 56
Mobility of persons and capital, 10, 30, 32
Monetary Council, 58
   isolation of, 71
Monetary union, 58
Monopoly, 30, 36, 48
Multilateral Free Trade Treaty, 23, 24, 30, 67

Negative effect of integration, 65
Newspapers, daily, 11
Nye, J. S., 17

OAS, 6, 62
Obstacles: to integration, 27
   to migration, 11
   program, 53-54, 72
   to tariff unification, 45-47
Oil and oil derivatives, 42-43, 45
Organization of American States (OAS), 6, 62
Organization of Central American States (ODECA), 16, 63, 72, 73, 74

failure of, 17
purposes of, 16

Payments, system of, 57
Peasants, 4, 64
Plan: general, 19-20
   trade liberalization, 23
Planning, 19, 62
Plans, for roads, 50-51
Political changes, results of, 14
Politicization, 85
   possibility for, 68, 70
Population, distribution of, 3, 7, 9-10
Prebisch, Raúl, 13-14, 18
Price support, 55
Problems, regional, diagnosis of, 13-14
Product specialization, 44
Products: excluded from free trade list, 24, 32, 42-44, 45, 55
   free trade list of, 23
   list of, for common tariff, 23-24, 45
Program of economic integration: acceleration of, 32, 34, 35, 38
   aims of, 41
   basis for, 21
   costs of, 20-21, 22, 24, 85
   financing, *see* Financing, Foreign aid goals of, 38, 47
   initiation of, 15-16
   obstacles to, 53-54, 72
   spill-over, 67, 73
   support from social groups, 75
Prosperity, period of, 2-3
Protection (industrial), 46-47, 48, 61, 80
   negotiation for, 76
Protocols, 55-56, 59
   tariff, 45
   See also Treaties
Public sector, post-war, 3, 4

Radio receivers, number of, 11
Railroads, 8, 50
Raw material, protection of, 46
   share of trade, 43
Reciprocity: defined, 77
   principle of, 19-21, 23, 85
Regime of Integration Industries, 23, 24, 30, 32, 34, 39, 67, 85
   failure of, 36-37, 47-48
   objectives of, 36, 47
   US opposition to, 36, 48
Regional bank, 37
Regional federation of chambers of industry (FECAICA), 76
Regional Office for Central America and

Panama Affairs (ROCAP), 37, 63, 74
Regional Telecommunications Commission (COMTELCA), 54
Relevance of integration, 64, 65
Requirements for US participation, 29–31
Requisites of the integration policy, 19–20
Revolution(s), 4–6
  Cuban, 28, 35–36, 39
Roads, construction of, 50–51

Sea ports, 8, 9, 53
SIECA, 51, 70, 71
  financing by, 62
  financing of, 74
  role, 69
  tasks, 70
Socarras, Prio, 6
Social structure, 3–4
Special Fund, UN, 74
Special System for the Promotion of Productive Activities, 48
Spill-over, program, 67, 73
Stabilization fund, proposed, 59
Stages of integration, 41–42
Status quo: agents of, 85
  reinforcement of, 80
Studies: balance of payments, 59
  by Central American Bank, 51
  communications, 54
  by ECLA, 21, 23
  monetary, 59
  roads, 51
  by SIECA, 59
  by UN, 9, 49–50
  transportation, 9, 49–50, 51
Subcommittees, role of, 21–22
Surcharge on import duties, 59

Tariff equalization, 23, 32, 35, 42, 45–47, 67
Tax holidays, use of, 46, 48, 61
Taxes: import and export, importance of, 3
  use of sales and consumption, 44, 59, 80
Technical assistance, source of, 61–62
Technocrats, 15, 17, 18, 70, 76–77, 79, 80, see also Técnicos
Técnicos, 5, 14, 17, 84–85, see also Technocrats
Telephones, number of, 11
Tourists, 11
Trade, 7, 29, 43–44
  quantitative restrictions on, 43, 55–56
  difficulties in liberalization of, 42–44
  See also Free trade
Trade Subcommittee, 23, 34
Transportation, 3, 8, 50–53
  air, 9, 51–53
  costs, 51
  railroad, 50
  road, 8, 50
  sea, 9
  studies, 50
Treaties, 6, 32, 34, 42, 54
  common external tariff, 24, 67
  free trade, 19, 23–24, 30, 67
Tripartite Treaty, 32, 34

United Fruit Company, 8
United Nations (UN): financial assistance from, 22, 60, 62, 63, 67, 74
  Security Council, 6
  technical assistance from, 21, 22, 24, 61, 83, 84
United Nations Technical Assistance Administration (UN-TAA), 21, 22, 24, 83, 84
United States: dualism of foreign policy, 39
  financial aid from, 8, 28, 32, 35, 37, 38, 39, 50, 60
  Latin American relations, 27–28
  participation of, 31–33
  policy shifts of, 28–29, 35
  private investment from, 61
  requisites for program participation by, 29, 31
  role of, 85
Urquidi, Victor, 16, 18

Values: of the elites, 3, 5
  revolutionary, 5

Wealth, distribution of, 3
Wheat and wheat products, 42–43, 45

About the Author

The author is a Guatemalan citizen. This study is his doctoral thesis for the Graduate Institute of International Studies of the University of Geneva, Switzerland. He has also written several articles on Central America in professional journals.